The Handbook of Sustainable Refurbishment

The Handbook of Sustainable Refurbishment

Non-Domestic Buildings

Nick V. Baker

Earthscan works with RIBA Publishing, part of the Royal Institute of British Architects, to promote best practice and quality professional guidance on sustainable architecture.

publishing for a sustainable future

London • Sterling, VA

First published by Earthscan in the UK and USA in 2009

ISBN: 978-1-84407-486-0

Typeset by FiSH Books, London
Cover design by Yvonne Booth
Graphics by Mike J.V. Baker

For a full list of publications please contact:

Earthscan
Dunstan House
14a St Cross St
London, EC1N 8XA, UK
Tel: +44 (0)20 7841 1930
Fax: +44 (0)20 7242 1474
Email: earthinfo@earthscan.co.uk
Web: **www.earthscan.co.uk**

22883 Quicksilver Drive, Sterling, VA 20166-2012, USA

Earthscan publishes in association with the International Institute for Environment and Development

A catalogue record for this book is available from the British Library

Library of Congress Cataloging-in-Publication Data
Baker, Nick (Nick Vashon)
The handbook of sustainable refurbishment : non-domestic buildings / Nick V. Baker.
p. cm.
Includes bibliographical references and index.
ISBN 978-1-84407-486-0 (hardback)
1. Buildings–Repair and reconstruction. 2. Public buildings–Repair and reconstruction.
3. Commercial buildings–Remodeling. 4. Sustainable buildings–Design and construction. I. Title.
TH3401.B35 2009
690'.24–dc22

2009007564

At Earthscan we strive to minimize our environmental impacts and carbon footprint through reducing waste, recycling and offsetting our CO_2 emissions, including those created through publication of this book. For more details of our environmental policy, see www.earthscan.co.uk.

This book was printed in Malta by Gutenberg Press.
The paper used is FSC certified and the inks are vegetable based.

Mixed Sources
Product group from well-managed forests, and other controlled sources
www.fsc.org Cert no. TT-CoC-002424
© 1996 Forest Stewardship Council

Contents

Preface xi

List of Acronyms and Abbreviations xii

Part One Principles

1 Strategy for Low Emission Refurbishment 3

1.1 The case for low emission refurbishment: Energy use in buildings 3
1.2 Refurbishment versus rebuild: Economics and environmental impact 3
1.3 The building, plant and occupants as a system 4
1.4 Implications for change of use 5
Impact on energy consumption 6
1.5 Environmental comfort standards 7
1.6 Passive environmental strategies 8
Natural ventilation 9
Daylighting 10
1.7 Prioritizing refurbishment options 13
Quantifying energy benefits 14
1.8 Integration with newbuild 18
1.9 Eco-communities and urban renewal 19
1.10 Environmental regulation 20
Energy Performance of Buildings Directive 20
Using other legislation in the UK 22
Voluntary schemes and drivers 22

Part Two Practice

2 Floors 27

2.1 Solid ground floors 27
Insulation options 27
Underfloor heating or cooling 27
2.2 Suspended ground floors 28
Insulation options 28
Underfloor heating or cooling 28

2.3 Intermediate floors 29
2.4 Thermal response implications of floor insulation 29

3 Walls 31
3.1 Solid walls 31
External insulation 31
Implications for external insulation 32
Internal insulation 33
Thermal response 33
Cold bridges 33
Interstitial condensation 34
3.2 Cavity walls 35
Insulation options 35
Practical considerations 36
Interstitial condensation 36
Thermal implications 37
Retrofit inner or outer leaf 37

4 Roofs 39
Roof types 39
4.1 Insulating roofs with attic spaces 40
Ventilation of attic space 40
4.2 Insulating roofs with voids 40
4.3 Insulating solid roofs 41
Insulation above the waterproof membrane 41
Insulation between waterproof membrane and structural deck 42
Insulation below the structural deck 43
4.4 Other thermal issues 43
Surface reflectance 43
Low-emissivity membranes in cavities 44
Thermal mass 44
Cold bridges 45
4.5 Green roofs and roof ponds 45
Green roofs 45
Roof ponds 46

5 Windows 47
5.1 Glazing materials 47
Heat transmission through glazing 48
Radiation transmission through glazing 50
High performance glazing 50
5.2 Framing and support systems 51
Obstruction of light due to framing 52

Thermal performance of framing 53
Framing material 54
5.3 Modifying apertures 55
5.4 Shading systems 56
Daylight redistribution 56
Shading options for refurbishment 57
External shading 57
Internal shading 59
5.5 High performance daylighting 60

6 **Atria and Double Skins 63**
6.1 Atria and energy: Principles 65
Thermal performance 65
Winter performance 65
Summer performance 67
6.2 Effect on daylighting 67
6.3 Planting and vegetation 68
6.4 Double skins and energy 68
6.5 Other environmental factors 68
6.6 Atria and double skins as part of sustainable refurbishment 69

7 **Mechanical Services and Controls 71**
7.1 Boilers 71
7.2 Heat distribution 72
Water 72
Air 72
7.3 Heat emitters 73
Positioning emitters 75
Sizing emitters 75
Coolth emitters 76
7.4 Fans and pumps 76
7.5 Refrigeration 77
7.6 Lighting installations 77
Luminous efficacy 77
Illuminance level and distribution 79
7.7 Controls 79
Local control 79
Central control 80
Zoning 80
7.8 Lighting controls 81
Occupancy detection 81

Daylight detection 82
Zoning 83
Energy savings 83

7.9 Building energy management systems 83

7. 10 Adaptive controls 84
Feedback 85
Caretaker controls 86

7.11 Hybrid and mixed mode systems 86

8 Renewable Energy Options 89

8.1 Other renewable energy technologies 89

Part Three Case Studies

9 The Albatros, Den Helder, The Netherlands 93

Objectives 93
Refurbishment strategy 93

The double skin 94
Performance of double skin 96

Ventilation and heating 96
Performance 97

Daylighting 98

Overall energy performance 98

Comfort 100

Conclusions 100

10 Lycée Chevrollier, Angers, France 101

Strategy for sustainable refurbishment 101

Main low energy measures 102
A. Thermal 102
B. Lighting 102
C. Comfort: Shading and ventilation 102
D. Other features 102

Insulation 102

Daylight and artificial lighting 103
Artificial lighting 103
Performance 104

Ventilation 104
Performance 104

The atrium 105

Photovoltaic panels 106

Waste management and other environmental issues 106
Overall energy performance 107
Gas consumption 107
Electricity consumption 107
CO_2 emissions 107
Comfort 108
Conclusions 109

11 Daneshill House, Stevenage, UK 111
Strategy for refurbishment 111
Main innovative energy-saving features 112
The CoolDeck system 112
Performance 113
Energy efficient air-conditioning controls 114
Performance 114
Energy-efficient lighting controls 114
Performance 115
Light emitting diode (LED) lighting in Customer Service Centre 116
Solar water heating array 118
Performance 118
Increased space use efficiency 119
Post occupancy evaluation 120
Overall energy performance 121

12 Ministry of Finance Offices, Athens 123
Refurbishment strategy 123
Main energy saving features 124
Fabric improvements 124
Night ventilation techniques 124
Ceiling fans 124
Daylighting and artificial lighting 124
Heating 124
Cooling 124
Ventilation 124
Energy management, control and monitoring 125
Performance 125
Thermal comfort and air quality 125
Daylighting and artificial lighting 127
Daylighting performance 127
The photovoltaic array 127
Overall energy performance 127
Heating 128

Cooling ... 128
Comfort surveys ... 128

13 The Meyer Hospital, Florence ... 131
Refurbishment strategy ... 131
The greenhouse ... 132
Daylighting ... 134
Overall energy performance ... 134
Comfort ... 134

Appendices ... 135

Index ... 165

Preface

In most European cities there is a vast stock of existing buildings, many of which are getting to the end of their useful life. To replace the stock would take several decades and incur an unrealistic financial burden. It would also create a large contribution to CO_2 emissions, as a result of the energy associated with the production of materials and the construction of replacement buildings.

It is therefore essential that we develop strategies and techniques to improve the energy performance of our existing stock. It is commonly understood that the heating, cooling, lighting and ventilation of buildings accounts for nearly half of global energy consumption, with the consequent CO_2 emissions having an effect on global warming. The reduction of day-to-day consumption of fossil fuels for heating, cooling, lighting and ventilation must be the main objective in any attempt to refurbish a building sustainably.

This guide is a product of the European Union (EU) funded REVIVAL project, which set out to demonstrate some of these principles by incorporating them in five refurbishment projects of large non-domestic buildings. Wherever possible it draws from the experience of the REVIVAL project, but includes other examples and illustrations when necessary.

This guide is aimed at the architect, engineer, surveyor and project manager. It sets out the case for sustainable refurbishment and the principle measures that can be adopted. It presents principles in a concise technical language, but follows with an explanation of practical implications. It does not attempt to be a source book of manufacturer's information and technical data, or to deal with construction detail.

REVIVAL Team
July 2009

List of Acronyms and Abbreviations

AC	air-conditioned
ANV	advanced natural ventilation
BEMS	building energymanagement systems
BREEAM	Building Research Establishment Environmental Assessment Method
CHP	combined heat and power
CRC	Carbon Reduction Commitment
COP	Coefficient of Performance
CSR	Corporate Social Responsibility
DEC	Display Energy Certificate
DEFRA	Department for Environment, Food and Rural Affairs
DF	daylight factor
DX	direct expansion
EEAS	Energy Efficiency Accreditation Scheme
EPBD	Energy Performance of Buildings Directive
EPC	Energy Performance Certificate
EU	European Union
IR	infrared
IRC	internally reflected component
LED	light emitting diode
l/w	luminaires per watt
MM	mixed mode
NV	natural ventilation
PAC	partially air-conditioned
PCM	phase change material
PIR	passive infrared
PSALi	permanent supplementary artificial lighting
PV	photovoltaic
SBS	sick building syndrome
SHF	Solar Heat Gain Factor
UV	ultraviolet

Part One **Principles**

1 Strategy for Low Emission Refurbishment

1.1 The case for low emission refurbishment: Energy use in buildings

In the non-domestic sector in Europe, building refurbishments offer far more opportunities for reducing emissions than new building; the latter represents annually less than 1.5 per cent of the building stock. The usual motivation for refurbishment includes:

- replacement of degraded finishes and components;
- tailoring space organization to new uses;
- improving environmental quality.

These reasons may be sufficient in themselves to justify the cost. If at the same time the building can be made more energy efficient, there will be a reduction in running cost and a reduction in CO_2 emissions. This will often be at a modest extra cost that can be justified by reduced running costs, or in some cases even, no extra cost.

1.2 Refurbishment versus rebuild: Economics and environmental impact

There are many instances when demolition and rebuild will be considered as an alternative to refurbishment. This could be justified purely on economic grounds, or the advantages offered by a new building could be considered to justify the extra cost. However, two non-economic factors should be considered:

1 The environmental impact of refurbishment versus newbuild.

2 The socio-economic impact.

Initially, the environmental impact of refurbishment will almost always be less than demolition and newbuild. This is because all the materials carry embodied energy – to replace them causes new carbon emissions (Figure 1.1). Furthermore, the demolition process and waste disposal creates carbon emission as well as other waste disposal impacts.

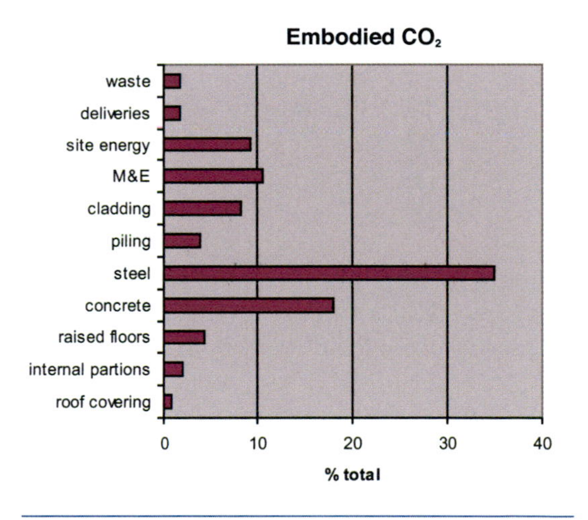

Figure 1.1 Embodied CO_2 associated with newbuild and refurbishment. Note large CO_2 content for bulk materials such as concrete and steel. Components made of these materials are the ones that are not normally replaced in refurbishment

Source: Thomas Lane quoting the Simons Construction Group in 'Our dark materials', *Building*, 9 November 2007

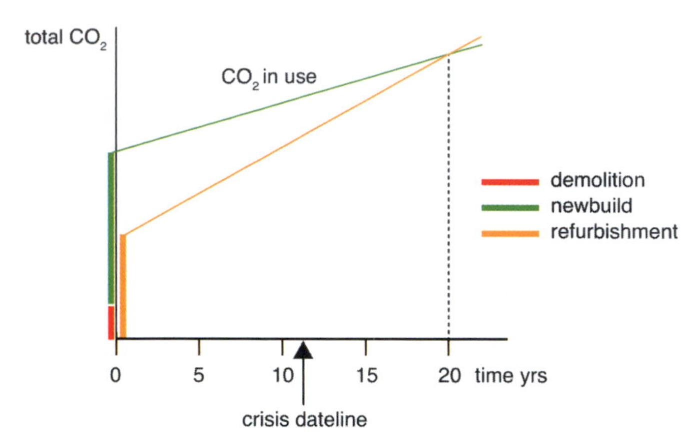

Figure 1.2 CO_2 emissions for newbuild and refurbishment as a function of time. The break-even point is very dependent upon the difference in performance (indicated by the slope of the graph) of the refurbished and newbuild.

It is often argued that a new building will operate at higher energy performance than a refurbished one, and that during its lifetime, may have less environmental impact. This dynamic relationship is shown in Figure 1.2. It demonstrates two important effects – that newbuild is only the lowest emitter after the break-even time period, and that this period can be extended by improved performance of the refurbished option. It also demonstrates that if the break-even time is beyond the time of the environmental crisis (or emission reduction target), the life-cycle emission is irrelevant and the refurbished building is the best choice. It is also evident that the break-even point is sensitive to the *actual* performance of the buildings; new buildings have not in general performed as well as predicted and this will postpone the break-even point.

The second consideration is about social benefit and employment. Generally, refurbishment carries a higher proportion of labour cost than newbuild. For example, the repair of a concrete structure and the cleaning of concrete finishes will direct money to tradesmen that in the case of new build would go to investors in concrete and steel manufacture.

1.3 The building, plant, and occupants as a system

Building simulations and analyses of monitored data have shown that the building fabric alone does not narrowly determine the energy performance. Figure 1.3 shows the performance being determined by three sub-systems, each having a variance in performance of about two-fold. When a poor building combines with badly designed systems and poor management, the resulting energy performance can be dramatically worse than the best. This wide variation of performance has been observed as shown in Figure 1.4. It is interesting to note that building no. 92 (extreme right) was built in 1987 and refurbished in 1992.

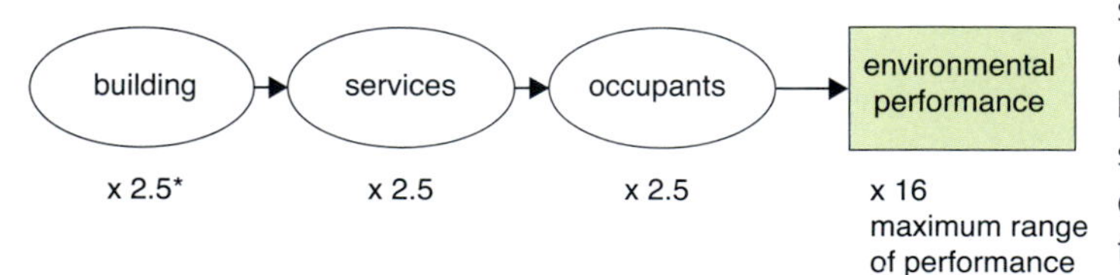

Figure 1.3 The building, the mechanical services and the occupants, as a system. Each controls a range of performance. A poor building may require much input from services which if badly managed leads to high energy consumption. The reverse may also be true. This accounts for a wide variance in energy consumption of similar buildings

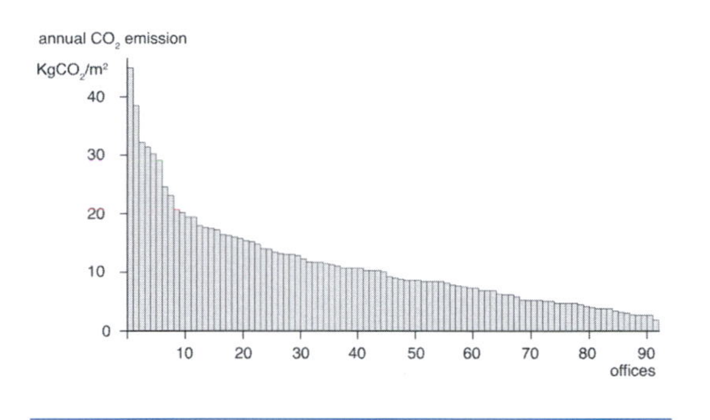

Figure 1.4 The annual CO_2 emissions per m^2 for 92 office buildings in the UK. The 20-fold variance illustrates the interactive effect between building, services and occupants

Source: Data from *Energy Consumption Guide 19 – Energy Efficiency in Offices* (1998) BRECSU, Watford

This evidence weakens the case for newbuild, since it shows that the inherent properties of the building are only one of the determining factors. This is particularly true in non-domestic buildings where overall energy consumption is dominated by processes and activities in the building. Both systems and management can be of as high performance in a refurbished building as in a newbuild.

Figure 1.5 Deep-plan buildings such as these, built in the era of cheap energy and relying on air-conditioning and artificial lighting, may present insurmountable problems for sustainable refurbishment

However, there may be individual cases where the inherent qualities of a building present insurmountable problems. For example, buildings with very deep plans, relying entirely on air-conditioning and artificial lighting, built in an era of cheap energy (Figure 1.5), will always be problematic both for their energy consumption, and internal environmental quality. Thus it is important to properly assess the potential for refurbishment, before conclusions are drawn.

1.4 Implications for change of use

Refurbishment is often accompanied by change of use. This may be across recognized use types – for example a nursing area of a hospital becoming an administrative centre (Figure 1.6), or a change from residential to office use (Figure 1.7). Or it may be that within a use type the functional demands on spaces are changing due to reorganization and the impact of changes in practice and technology. For example, developments in IT have a continuing influence on office practice and the spaces that support it.

Figure 1.6 REVIVAL building Meyer Hospital, Florence: Refurbishment accompanies change of use from nursing area to hospital administrative and reception area, involving different environmental conditions

Figure 1.7 REVIVAL building The Albatros, Den Helder, The Netherlands: Refurbished in conjunction with change of use from residential to offices

It is difficult to generalize here, but it could be said that opportunities are sometimes missed because designers impose stereotypical solutions, often ignoring the serendipity of fitting a new function into a building generated by a different set of aims.

For example, in a conversion of an old factory workshop to modern office use, high ceilings with exposed structural slabs are often replaced with suspended ceilings for acoustic reasons, and under the misguided impression that the original spaces would be impossible to heat efficiently. This action not only destroys much of the architectural quality of the space, but will also have a negative influence on daylight distribution, natural ventilation and, possibly, thermal response.

Change of use may bring about changes in purely technical parameters. These include:

- occupancy pattern and density;
- internal gains;
- lighting levels;
- ventilation rates;
- thermal set-points and response;
- acoustic properties (reverberation time, noise exclusion).

These changes may bring about benefits and disbenefits. For example, an historic warehouse converted into a library will create a difficult challenge to the designer if the intention is to provide daylight, due to the shallow floor-to-ceiling height. On the other hand, a heavyweight building that required wasteful intermittent heating in its original function as a primary school, would not be so inefficient if used for a much longer occupied period as, for example, a health clinic. Furthermore, the intermittent heating would be less wasteful anyway if the envelope insulation was improved as part of the refurbishment package.

Thus, the inherent properties of the building, the operational requirements of the new use, and the technical options in the refurbishment all have to be considered interactively.

Impact on energy consumption

In spite of improvements to the performance of the fabric and systems, change of use may bring about an increase in the energy consumption. This does not necessarily mean that the low-energy refurbishment has failed, since the measures adopted have undoubtedly led to lower energy consumption than if absent.

In measuring the success of the refurbishment then, it would be fair to make a comparison of the building's actual energy performance (shown at A in Figure 1.8) with, firstly, the existing building under the new use and complying with accepted comfort conditions, but without adopting low-energy measures (B). Secondly, a comparison should be made with a new building of similar use type (C). We might expect a performance somewhere between these two or with really successful refurbishments, even surpassing typical newbuild performance. Finally, a comparison should be made with the average emissions for the building stock of the same use type (D).

The example here shows that the refurbished building emits 35 per cent of that predicted for the original building with a change of use, and

58 per cent of that of the measured average for the existing building stock, although it emits 10 per cent more than a new building. The fact that the refurbished building with its new use emits nearly twice that of the original building with its original use, is not of much relevance.

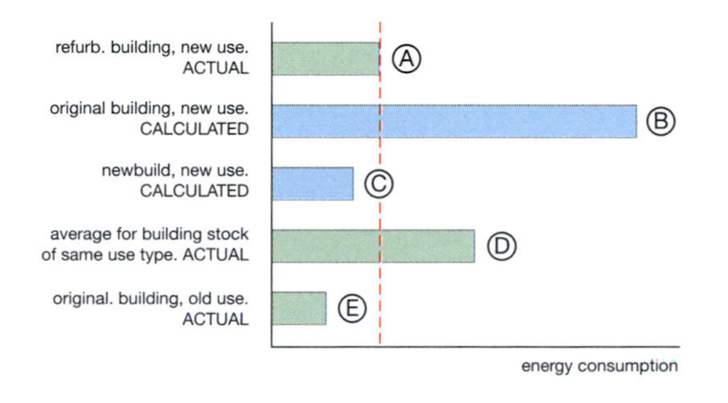

Figure 1.8 Comparing like with like: Assessing the improved performance of a refurbished building

1.5 Environmental comfort standards

Improved comfort standards are often the initial motivation for refurbishment. The building shown in Figure 1.9, an office block built in the 1970s, is poorly insulated with large areas of single tinted glazing, no shading and a poorly controlled heating system. Since its original occupation, density has increased, and there has been a proliferation of computers and other business machines. The frequent complaint is overheating in summer, both under-heating and overheating in winter, and poor air quality.

The client's presumption may well be that full air-conditioning would be the answer. If conventional comfort standards were sought, this could indeed true, although this would be neither an economical nor environmentally friendly solution.

If refurbishment measures included shading, improving the envelope insulation, and reinstating openable windows, comfort conditions would be greatly improved, although the strict

Figure 1.9 An over-glazed lightweight building of the 1970s suffering from overheating in summer and under-heating in winter. The client's perception is often that the only solution to comfort problems is air-conditioning

standards achievable by air-conditioning may still not be met for all of the year. However, it is now widely accepted that for buildings running under predominantly passive systems, occupant satisfaction can be high, even when the conventional standards are not met.

A key factor is the presence of *adaptive opportunity*. This is the ability of the occupant to make changes to the environment, and/or make changes to their personal condition, in order to improve their comfort. Typical opportunities that might be present are listed below.

Positive adaptive attributes

- relaxed dress code;
- occupant mobility;
- access to hot/cold drinks;
- openable windows;
- adjustable blinds;
- desk fan or locally controlled ceiling fan;
- local heating/cooling controls;
- workstation/furniture flexibility;
- shallow plan (minimizing distance from windows);
- cellular rooms (reduces mutual disturbance);
- surface finishes appropriate to visual task;

- daylight and task lighting backup;
- good views (external and internal);
- transitional spaces (verandahs, atria, etc.);
- good access to outside areas.

Negative adaptive attributes

- uniformity of physical environment (temperature, lighting, colour);
- deep plan, reduced access to perimeter;
- dense occupation with restricted workstation options;
- sealed windows;
- views obstructed by fixed shading devices;
- central mechanical services control.

Studies have shown (Baker and Standeven, 1994) that the presence of several of the positive attributes will result in occupants tolerating temperature excursions typically up to 5°C above conventional upper temperature limits, and around 3°C below conventional lower limits. This may allow the designer to opt for the passive solution rather than the air-conditioned solution.

This will have implications for initial cost, maintenance cost and carbon emissions. It will also have implications for the choice and prioritizing of refurbishment measures. For example, if it were decided to air-condition the building, the replacement of standard double glazing with high performance low-e units would have a greater impact on carbon emissions than if the building were to be freely ventilated by openable windows. This is because the temperature differential in the latter case would be small or even non-existent. However in *both* cases, shading would be highly beneficial.

In some cases, exceptional overheating conditions may be unacceptable, although conditions prevailing for most of the year may be satisfactory without air-conditioning. In this case, intermittent comfort cooling may be applied. The technological aspects of this are discussed in section 7.11. Here we make the point that a modest dependence on comfort cooling may result in a building being able to take the predominantly passive option. Furthermore, because of the intermittent nature of the comfort cooling, and its controllability, it will not be necessary to apply such strict comfort limits as in a conventional air-conditioned building. This strategy, often referred to as 'hybrid' or 'mixed mode', results in comfort cooling often being a viable and energy efficient option.

1.6 Passive environmental strategies

Statistically, air-conditioned buildings consume significantly more energy than naturally ventilated buildings. In temperate climates, field studies have shown that in spite of the extra capital and running costs, occupant satisfaction was no greater than in naturally ventilated buildings (Figure 1.10). Even in hotter climates, as a study of office buildings in Lisbon showed, satisfaction in some air-conditioned buildings may be significantly less than in some naturally ventilated buildings. Thus the strategy for avoiding air-conditioning is a good one, although hybrid systems and comfort cooling (described later in section 7.11) may represent a viable alternative.

The situation often faced in refurbishment is of a building with very poor comfort conditions, where air-conditioning is seen as the only solution. However, it could be that after making fabric and system improvements, comfort conditions become acceptable. This should be tested by analysis or simulation before the air-conditioning option is adopted.

This may even apply to a building that has air-conditioning already. Many over-glazed buildings of the 1960s and 1970s were subsequently air-conditioned to make conditions bearable. However, measures such as shading, fabric insulation, reduction of glazing area, adoption of adaptive controls, may well render full air-conditioning unnecessary. Even if air-conditioning is adopted, these measures will reduce the air-conditioning loads significantly.

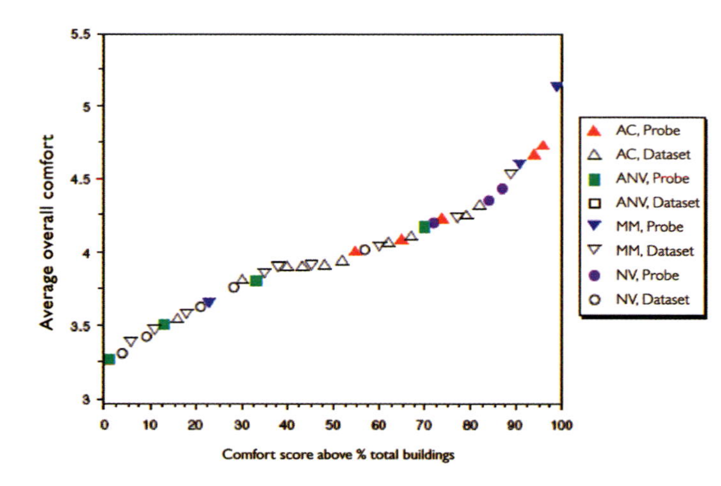

Figure 1.10 Overall occupant comfort by ventilation types; natural ventilation (NV), advanced natural ventilation (ANV), mixed mode (MM) and air-conditioned (AC). The coloured plots are for the PROBE (UK) survey and the open plots from the Building Use Studies data. Summer survey results from 26 office buildings in Lisbon show that occupant satisfaction in naturally ventilated buildings (NV) and partially air-conditioned buildings (PAC) was higher than in air-conditioned buildings (AC)

Source: The Probe Study (2001) *Building Research and Information*, vol 29, 2 March

Natural ventilation

Many candidate buildings for refurbishment will have high rates of infiltration. This is particularly true of buildings constructed using pre-cast concrete components, and panel and curtain wall systems, with dry linings – typical construction techniques of the 1960s to 1980s era (Figure 1.11). In buildings with original glazing systems, these too were very leaky. In cooler climates, the result of the high infiltration rate is a waste of energy due to heat loss. However, one benefit of the high air-change rate was good air quality.

Figure 1.11 REVIVAL building Lycée Chevrollier, Angers, France: Built in 1958 of pre-cast concrete construction with dry lining and poorly fitting metal windows, it had very high infiltration rates before refurbishment

Refurbishment measures to reduce uncontrolled infiltration must then recognize the need to provide ventilation to ensure that a minimum air quality is maintained. This does not mean that the benefits of the more airtight envelope will all be lost, provided some means to prevent over-ventilation is present. Broadly, the principle is 'build tight, ventilate right'.

In predominantly warm climates (i.e. with little winter heating) reduction of infiltration only brings energy benefits if the building is air-conditioned to a temperature significantly below the prevailing outdoor temperature. If it is not to be air-conditioned the main concern switches to getting enough ventilation to remove heat gains and to generate air movement. This will demand large openable areas.

Night ventilation

A further important function may be the provision of night ventilation to cool the structural mass. The principle is illustrated in Figure 1.12. This function will also require large openable areas in the envelope, and unobstructed flow paths within the building. Furthermore, it is essential that the ventilating air can be thermally coupled with the thermal mass of the building. These requirements may present design challenges, particularly in relation to security and noise control.

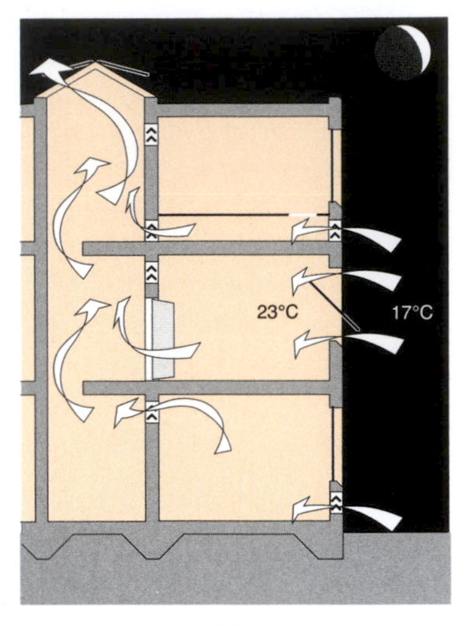

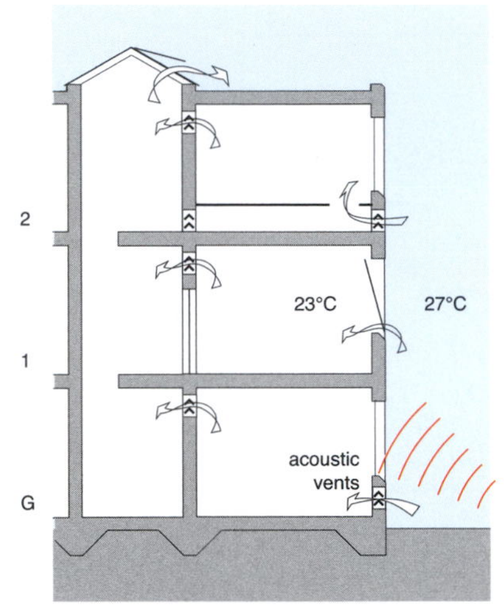

Figure 1.12 The principle of night ventilation: The mass of the building is cooled at night to provide a heat sink for internal gains during the day

Daylighting

For the two or more centuries since the industrial revolution, when man moved indoors, daylight had been too valuable an asset to waste. Architects responded by a whole typology of window and rooflight designs, readily accepting the need for shallow plans, light wells and courtyards (Figures 1.13, 1.14). This constraint was reinforced by the same need for shallow plan to achieve good natural ventilation. Whilst, of course, artificial light had to be employed after the hours of darkness, daylight was always the preferred source in the daytime, and artificial light was regarded as inferior both technically and on health grounds. The latter dated from pre-electric times when oil and gas lighting created serious levels of indoor pollution and fire risk.

The development in the 1950s of the fluorescent lamp, together with relatively cheap electric energy, prompted designers to question the need to provide the working illumination by daylight (Figure 1.15). This freed them to adopt deeper plans, which in principle made economies in space efficiency and building cost.

Figure 1.13 Wauquez Department Store (now a museum of cartoon art), Brussels: Rooflighting the top floor (1920). The architect Horta could not resist architectural elaboration

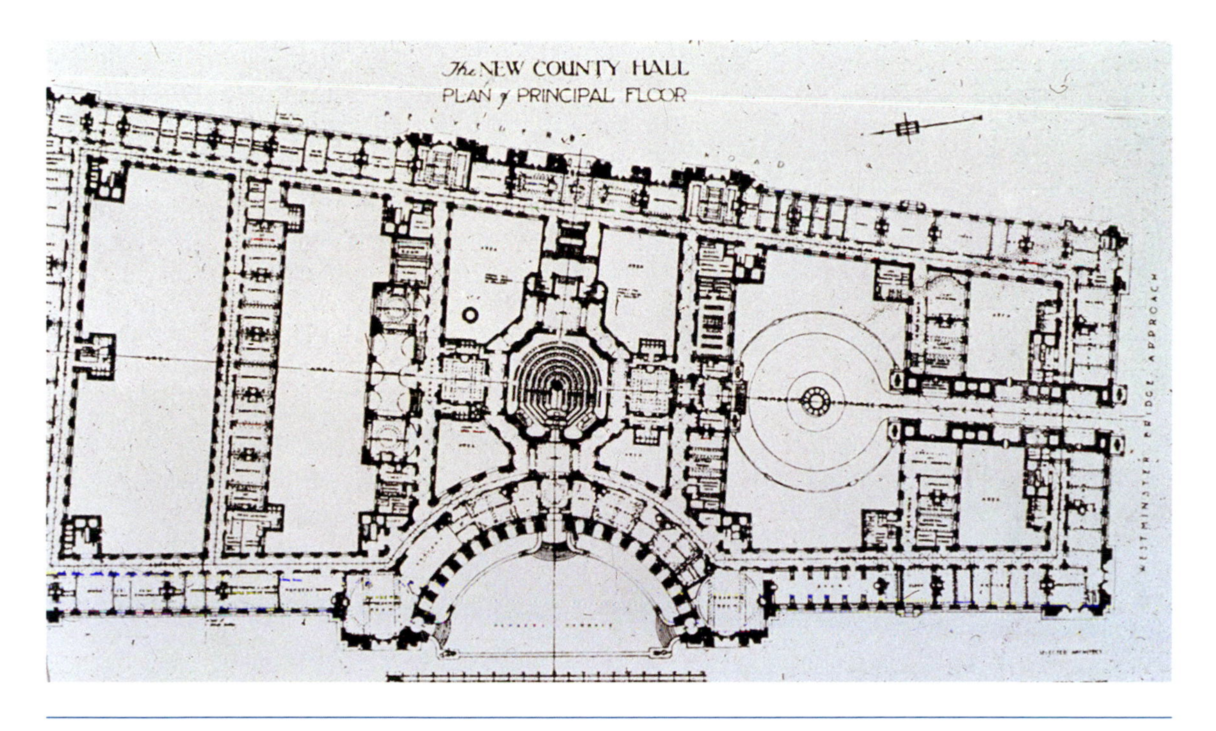

Figure 1.14 County Hall, London, 1908: Even large buildings were in fact shallow plan, providing access to daylight and natural ventilation

Buildings of the 1960s to 1980s found themselves at a crossroads. On the one hand glass and glass-supporting cladding technology had made great strides to support the modernist, minimalist view of facade design, often with huge areas of glazing (Figure 1.16). On the other hand, the improvement by a factor of five in the luminous efficacy (lumens light output per Watt electrical input) of the fluorescent lamp over the incandescent lamp meant that it was now technically and economically feasible to provide the working illumination entirely artificially.

The result of these conflicting influences meant that the art of good daylight design was virtually lost. Initially the over-glazed buildings were, not surprisingly, over-illuminated leading to glare, overheating in summer due to solar gain, and suffering from large heat losses in winter. Response to these included, ironically, permanent supplementary artificial lighting (PSALi) where the poor light distribution due to large areas of side-lighting was balanced by artificial lighting away from the perimeter zones. The next response was to reduce the light transmittance of the windows to values as low as 25 per cent, with tinted and reflective glass (Figure 1.17). Finally the contribution of daylight was abandoned altogether. This allowed a retreat into deep plan, and for a brief period a knee-jerk response of drastically reducing glazing areas. The resulting deep plan, air-conditioned, artificially lit buildings from this era have proved to be the highest energy users and the most likely candidates for sick building syndrome (SBS) (Figure 1.18).

Within our target building group all of these types will be found. The following Table 1.1 provides a response strategy.

Finally, the benefits of daylighting and good window design extend beyond the saving of energy. There is growing evidence that the view

Figure 1.15 The leap forward in luminous efficacy of the fluorescent lamp heralded the daylight dark ages. Deep-plan buildings abandoned daylight and natural ventilation, and windows were relegated to providing occasional views

Figure 1.16 The transparent building – huge areas of glazing and shallow plan: Delft University, circa 1960

from windows and the perception of the presence of daylight, even without direct views, is valued by occupants. This can lead to increased well-being and productivity, and also increased tolerance of non-neutral environmental conditions. The latter offers significant support to the adoption of a passive strategy.

Figure 1.17 First efforts to overcome the effect of large areas was to reduce the transmission of glass by reflective and absorptive glasses

Figure 1.18 The Kalamazoo building, Birmingham, completely abandoned daylight with deep plan, slit windows and full air-conditioning. Buildings of this type are prime candidates for sick building syndrome. It was demolished in 2003.

1.7 Prioritizing refurbishment options

Most building projects, whether refurbishment or newbuild, are ultimately budget limited. Budgets will constrain both sustainability measures and functional design solutions. In the former case, a typical example might be the installation of a photovoltaic array technically possible to meet the equivalent electrical load of the building (we will call this *environmental* benefit), but prevented by prohibitive cost. In the second example, a more generous budget for furnishings would be regarded as desirable but not viable (we will call this *personal* benefit). In some cases, budget constraints may interact with decisions which affect both of these areas – for example, limiting the space provision per occupant will probably be seen as a reduction in quality from the occupant's point of view, but may well improve the energy use per occupant. In other cases, a measure may create benefits in both categories – for example, improving access to daylight will reduce energy consumption and be seen as a positive move by most occupants.

The interaction of types of benefit (i.e. environmental or personal) and the sometimes conflicting, sometimes mutually supporting, outcomes, makes the prioritizing of refurbishment measures difficult. Consider the case where two measures – insulating the roof or installing shading devices – are of similar cost, but have different levels of impact on the occupants. Roof insulation will bring economic benefit to the building operator, environmental benefit due to reduced heating load, but no personal benefits. On the other hand, shading devices will improve personal comfort conditions, may improve the appearance of the building, and if the building is to be air-conditioned, may reduce cooling loads.

How do we resolve this question? We suggest that the environmental benefits and personal benefits have to be treated separately, initially. When there is a clear quantitative ranking of environmental measures, then personal benefits

Table 1.1 Daylighting strategies for refurbishment

Building type	Daylight status	Refurbishment strategy
Shallow plan, small glazing area	Daylighting provision degraded.[1]	Reinstate daylighting, install photo-responsive controls.
Shallow plan, large glazing area	Daylighting degraded by low transmissivity glass or fixed shading, or nothing.	Consider reduction of glazing area, high performance daylighting[2] (adjustable shading and high performance glass). Install photo-responsive controls.
Deep plan, large glazing area	Daylighting abandoned even in perimeter zone, as above. High levels of uniform artificial light.	As above for perimeter zone. Consider advanced daylight options to increase daylight penetration. Install rooflights (top floor) and consider light wells. Install task lighting with low illuminance photo-responsive background lighting.
Deep plan, small glazing area	Daylight abandoned. High levels of uniform artificial lighting.	Consider increasing glazing area together with measures above.

1 Daylighting performance may be degraded by inappropriate shading design, internal or external obstructions, poor distribution due to lowered ceilings and low reflectance surfaces, deliberate reduction of transmission of glass by films or paintwork.

2 The term *high performance daylighting* refers to the technical means to balance the usefulness of daylight to perform the visual task, and the positive benefits of view, with the disbenefits of overheating and glare (see 'High performance daylighting' Appendix).

should be considered. This is not because the personal benefits are any less important, but because they are far less easy to quantify. There may be cases, however, where the personal benefits are the primary concern, or may even be essential – for example, the replacement of asbestos with a non-toxic insulant.

Table 1.2 below is indicative only. All of the benefits are dependent on circumstances – for instance, the impact of improving plant efficiency obviously depends on the existing efficiency and the final efficiency.

Quantifying energy benefits

Here we review typical refurbishment options and apply them to a range of scenarios. Of course it is not possible to create global ranking of measures since their performance is very sensitive to the context. For example, improving envelope insulation to a building in a cold climate will have far greater benefit than in a mild climate, whereas the case for shading devices might be reversed. Similarly the building type will influence priorities – a shallow-plan school building occupied predominantly in the day will benefit far more from the investment of improved daylight access than an institutional building occupied for 18 hours per day. There are too many combinations of parameters to attempt comprehensiveness; rather, this example is given as an illustration of the interaction of measures and the strong impact on priority.

To assess these scenarios we have used the LT Europe software. This software developed from the LT Method takes account of the interaction between lighting heat gains, solar gains, occupant gains and losses through the envelope elements and ventilation. It evaluates energy inputs for heating, lighting, ventilation and cooling, and can indicate comfort levels resulting from the omission of mechanical cooling. It can also evaluate

Table 1.2 Environmental and personal benefits of refurbishment measures

Technical measure	Potential *environmental* benefit	Potential *personal* benefit
Change fuel type	large	zero
Improve plant efficiency	large	zero
Improve controls	large	moderate
Insulate envelope	large	zero to small
Improve daylight access	moderate	moderate
Install shading	moderate	moderate to large
Install task lighting	moderate	large
Increase occupant density	moderate	-ve small
Improve noise control	zero	large
Improve art. light spectral qual.	-ve small	small
Reduce occupant density	-ve moderate	moderate to large
Provide comfort cooling	-ve moderate	large

Note: Environment refers to global environment.

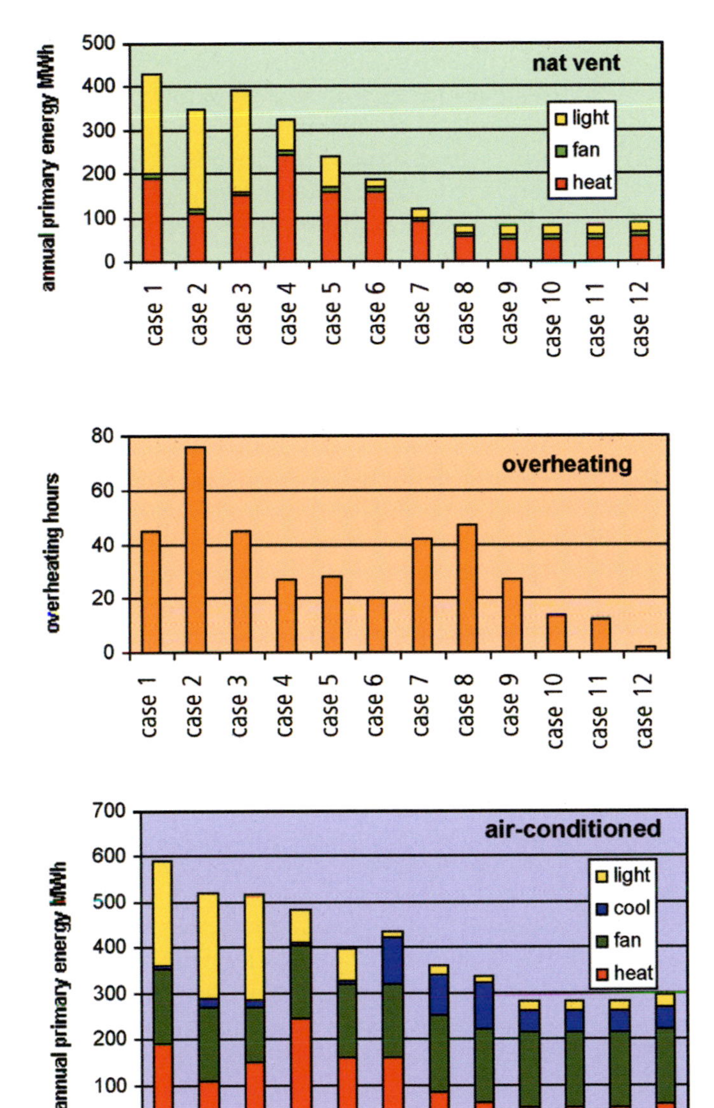

Figure 1.19 Case comparison from the LT Europe software showing the progressive improvement in energy and overheating performance for natural ventilation and air-conditioning

the impact of thermal mass and the benefit of night ventilation

The software computes the energy in primary energy units, which closely relate to CO_2 emissions. The cooling loads are calculated on a 25°C set-point. In the naturally ventilated case, overheating is defined as the number of days with two consecutive hours or more above 27°C.

The example studied here is a typical 1970s medium-sized four-storey office block with a 14m × 34m footprint. It is set in an open site in the UK, where there is both a significant heating and cooling load. The building is glazed on both long facades that face north and south, with single glazing occupying 70 per cent of the facade area. The envelope is leaky and poorly insulated (U-values around 1.5). Due to lack of solar controls, and internal gains due to low efficiency lighting and fast growing IT installations, frequent overheating has prompted the installation of air-conditioning, together with tinted window film. Controls are poor, and the centrally switched lighting is on all day. Occupant satisfaction is very low, energy costs high, and it is now in dire need of refurbishment.

Figure 1.19 shows the case comparison, displaying the total primary energy consumption for the base case described above (labelled case 1) followed by the application of 11 measures aimed at improving energy performance. This shows firstly how the measures interact, and secondly provides some evidence as to how their cost effectiveness could be ranked, although the cost of the measures is not given explicitly.

The cases explored are as follows:

Case 1	base case
Case 2	base case + low-e glazing
Case 3	base case + boiler efficiency increased from 65 per cent to 85 per cent
Case 4	base case + high efficiency lighting
Case 5	(4) + vent rate reduced 1.5ac/h to 1.0ac/h, boiler efficiency increased from 65 per cent to 85 per cent
Case 6	(5) + photoelectric lighting controls
Case 7	(6) + low-e glazing
Case 8	(7) + insulated opaque envelope
Case 9	(8) + reduced glazing area
Case 10	(9) + summer high vent
Case 11	(10) + night ventilation
Case 12	(11) + high thermal mass

Case 1. The building starts with high heating, lighting and cooling loads. Without air-conditioning, overheating is around 45 days per year, about ten times an acceptable level. Clearly the air-conditioning was essential.

Case 2. Here the envelope is identified as having poor thermal performance and it is proposed to reglaze with low-e double glazing. This is an expensive option and the results are disappointing with only a 10 per cent reduction in energy.

Case 3. This shows a much cheaper option of increasing the boiler efficiency from 0.65 to 0.85 per cent which shows a saving of 8 per cent. In reality, this may reduce emissions further due to improved combustion. It will also cause much less disturbance to the use of the building.

Case 4. This returns to the single glazing to concentrate on the other large component, lighting. The existing lighting situation is bad, with poor efficiency lamps and luminaires only delivering 20 useful lumens per watt (l/w)[1] with a design luminance of 300 lux. This can easily be improved to 40l/w and the design luminance reduced to 200 lux, made possible due to the widespread use of self luminous computer screens, and task lighting where necessary. This halves the lighting energy. But there is a penalty to pay in heating, and about half of the decrease of energy is lost. However the net effect for the air-con case is a reduction of 21 per cent, and 30 per cent for the naturally ventilated case, and is thus, together with the increased boiler efficiency, certain to be far more cost effective than the low-e glazing. Note too, that there is a significant reduction in overheating days, which had actually increased with the low-e glazing. However, overheating days are still around 28, which is unacceptable.

Case 5. The increase in heating load prompts two measures to improve the thermal performance. One is to improve the airtightness of the envelope, reducing the infiltration from 1.5 to 1.0 air changes per hour (ac/h), which may prove to be a technical challenge, and secondly to improve the heating plant further to an efficiency of 0.9, achievable with a condensing boiler. This largely compensates for the loss of heat gains from the lighting.

Case 6. Lighting energy is reduced further by the installation of photo-sensitive controls. As can be expected for a shallow-plan building with (more than) adequate glazing area, a large saving is made, reducing the lighting energy to about 35 per cent of the uncontrolled value. It is worth noting, that if this had been applied before the efficiency of the lighting had been improved, the absolute savings for this measure, and thus its cost effectiveness, would be even greater. In this case the reduction in lighting energy has only led to an increase in heating energy of about 10 per cent, and this is probably because the lights are going off in the middle part of the day, when for most of the year net heating loads are very small. However, together with case 4 it does demon-

1 This includes the lamp efficacy and the utilization factor of the luminaire and room combination.

strate the dependence of heating loads on casual gains.

Overheating has steadily reduced, down to 20 days, and there is now a real possibility of abandoning air-conditioning. So from now on, we will concentrate on the natural ventilation option.

Cases 7 and 8. We now return to test the effects of improvements to the envelope, and in case 7 low-e glazing is installed again. This results in a 40 per cent (of case 6) reduction in the heating load. When combined with insulation to the walls and roof (case 8) the reduction in heating load is 64 per cent (of case 6). This is of course a very expensive measure, and still could be of relatively poor cost effectiveness. However, if there was already a need for fabric improvements due to failure in weathering function, much of this cost would be offset.

Case 9. In case 9 the glazing area is reduced from 70 to 35 per cent on both south and north facades. This has little effect on energy balance – there is only a small increase in lighting energy, and the reduction of heat loss from the north facade is compensated by a loss of useful heat gains on the south facade. However, this strategy carries two important advantages. Firstly it reduces unwanted solar gains, significantly reducing overheating; secondly, it reduces the cost of low-e glazing; and thirdly, if expensive shading treatment is necessary, it will be needed on a much smaller area. It should be noted that the reduction would have had a much bigger impact on heat loads had the glazing not already been upgraded to low-e.

Case 10. However, the improved thermal performance of the envelope has had a downside with a disappointing increase in overheating due to the reduced heat losses (case 8), only partially compensated by the reduced glazing area. How do we increase heat losses on demand? By opening windows. When air-conditioning had been installed, the windows were sealed. By reinstating openable windows, high air-change rates are possible in summer, and overheating can be further reduced as in case 10.

Case 11. But overheating is still around 14 days per year. Night ventilation is known to be effective in cooling the structure at night with the cooler night air, and preparing the thermal mass to absorb gains made in the daytime. Case 11 with night ventilation shows a significant reduction to seven days.

Case 12. Here an attempt to make further improvements is tested, where during the fit-out, structural mass is exposed by removing suspended ceilings and carpeted floors. This reduces overheating to four days, but at the expense of some extra heating due to intermittent occupation.

Conclusions

The series of cases shows the interaction between measures and their varying impact. Although generic costs are not provided explicitly, it is quite clear that the measures will vary widely in cost effectiveness, and this should be borne in mind when establishing the priority of upgrading measures.

It also demonstrates that a critical point in temperate climates is the ability to move from air-conditioning to natural ventilation. This is partly because air-conditioning carries a large overhead of fan and pump energy, which is not directly proportionate to the cooling (or heating) load. The result, in this case, of moving from the air-conditioned base case, to the naturally ventilated final case is a reduction in primary energy consumption of 81 per cent. It must be pointed out, however, that improvements could have been made in the efficiency of the air-conditioning system that are not tested here, and which would have reduced the final difference.

Although some of the discussion has referred to percentage changes, in order to indicate impact, ultimately it is the absolute reduction of energy or CO_2 emissions against which a measure should be judged. Thus a further large percentage reduction to a load that has already been reduced to a small value, is less cost effective. This implies that the order that measures are applied is important.

Climate will affect the relative impact of measures. In warmer and sunnier climates, insulation will have less impact in the naturally ventilated case, and measures to reduce overheating more. Similarly, building type will influence the relative impact of measures. A building such as a junior school occupied for the middle part of the day will benefit more from improvements to daylighting than, say, a hospital, occupied for 24 hours per day, whereas in the case of thermal improvements to the fabric, the reverse may be true.

1.8 Integration with newbuild

Refurbishment projects often include varying degrees of newbuild (Figure 1.20). This could range from a small wing or extension, to major separate blocks totalling a built surface as great or greater than the original building. In some cases this may involve demolition and replacement, and in others, the newbuild may simply be intensifying the site coverage.

Figure 1.20 The new wing and atrium built as part of the Lycée Chevrollier (Angers, France) refurbishment. The old wing is on the right behind the retrofit shading

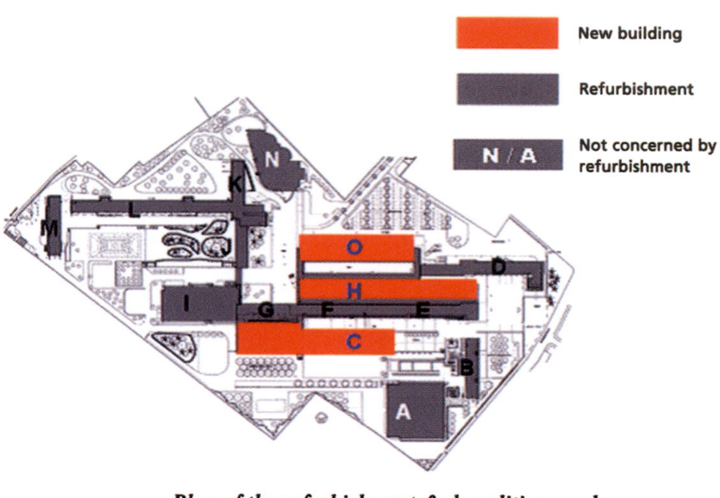

Figure 1.21 Plan of Lycée Chevrollier showing the new block N which caused some obstruction to daylight available in the original block H

We can identify two broad categories of integration that could affect environmental performance.

1 architectural
2 engineering.

Architectural integration includes consideration for the massing of the final building as well as the aesthetic and stylistic qualities of both old and new. Only the former issue is dealt with here.

The massing of a complex building or group of buildings can have a significant effect on the availability of daylight, the penetration of sunlight and the microclimate of spaces around and enclosed by the buildings. For example, on a visit in spring to the Lycée Chevrollier building, it was observed that the classrooms in block F had lights on, whilst most in block E were off. This clearly is a result of the obstruction of the new block C (Figure 1.21). This new block will also affect availability of lower angle sun, and thus may reduce existing overheating problems, and hence the need for shading.

This illustrates the need to give full consideration to the environmental impacts of the new massing and interaction of all parts of the project. Another REVIVAL example is shown in the Meyer Hospital Case Study (Chapter 12) where the addition of a new greenhouse on the south elevation affected the daylighting of the adjacent office rooms in the original building.

Engineering integration is concerned mainly with the integration of services. In particular, heating plant may serve all parts of the project, and according to the amount of increase of built surface area, the plant capacity may have to be adjusted. On the other hand it could be that existing plant will be adequate for an increased built area due to the savings made by the improvement to the existing refurbished part. In any event, unless the heating plant has been renewed recently, and the distribution mains are of a good standard, new plant, properly sized for the total development, is likely to be cost effective.

One further consideration, where the project is an extensive building complex, is that distributed plant may be the best option. This is because distribution mains for heating and cooling are costly in capital, maintenance and running costs, and with modern IT control and building energy management systems, personal physical presence to control plant in a central location is far less necessary than it used to be.

1.9 Eco-communities and urban renewal

This is a planning concept concerning community-scale policy and design rather than individual buildings. The term covers a wide range of principles, from near-spiritual attitudes about 'living with nature', to more pragmatic issues such as waste recycling and low carbon transport systems.

The prefix 'eco' (from the Greek *echos* – home), is rather widely used, but refers here to the concept of interdependence of buildings, the activities within them, and their occupants, in some synergistic way, that also supports sustainability and minimizes negative impacts on the wider environment. It is essentially opportunistic – just as in *ecology*, in the conventional biological sense, we see different organisms occupying niches in a continuous cycle of energy and material, so too in this more human-dominated context we see activities and land use having a similar interdependence locally, and thereby reducing global dependence.

There are three reasons why it is of relevance in refurbishment projects. Firstly, the project may be quite extensive – particularly if it includes housing where many units may be refurbished by the same principles. Clearly there are opportunities here to consider the area as a whole, how it interacts within itself and how it interacts with its immediate surroundings.

Secondly, some building types naturally interact with their surroundings both socially and physically. For example, a refurbished factory will need a workforce which in turn will require transport and/or accommodation. It may also produce low-grade heat as a by-product, which could be used for space heating via a district heating system.

Finally, it is particularly relevant in urban renewal areas, where the refurbishment may cover groups of buildings of differing use types, as well

Figure 1.22
Eco-community renewal scheme at New Islington, UK

Table 1.3 Potential for eco-interaction in large scale renewal projects

ENERGY SUPPLY	
Heat	Group solar thermal Waste industrial heat Heat from sewage and waste water Heat from waste incineration Combined heat and power Group solar thermal Waste high-grade biomass Biomass production from crops Ambient sources Geothermal (and cool) Marine (and cool)
Electricity	Combined heat and power Shared wind power Shared solar thermal with storage Shared PV arrays Hydro power Tidal power
ENERGY CONSERVATION AND COMFORT	
Microclimate	Wind sheltering with vegetation Wind sheltering by buildings Wind access in summer Solar shading with vegetation Solar shading by buildings Solar access in winter
Building specification	Maintenance Technical support
Outdoor amenity space	Microclimate Accessibility Adaptive opportunity Privacy Security Management Gardens and food production
Transport	Access to public transport network Cycle paths Cycle storage and security Footways Car parking and access Transport sharing schemes Traffic management hierarchy
Water and sewage	Rain water catchment Greywater management Reedbed purification
Waste disposal and recycling	Waste sorting Compost Re-use

as infrastructure, and changes of land use (Figure 1.22).

In the example above, this will have implications for decisions made in the refurbishment of both the housing and the factory, at a strategic and detailed technical design level. At a strategic level, the decision to use waste heat has a political aspect, since it will affect the cost both in money terms and carbon emissions, and will require management, and should bring benefit to both the user and the producer. At a technical level the use of waste heat rather than individual heat sources may require heat emitters that can function at low water temperatures, and externally will require consideration for the routing of the heating mains.

To the left is a table of the most likely possibilities for eco-interaction in relation to refurbishment projects. Most are concerned with energy production, but some are concerned with energy and resource conservation.

1.10 Environmental regulation

Building refurbishment is increasingly being brought under mandatory control in Europe and this is a trend that is likely to continue. The European Commission is driving forward various initiatives with which individual Member States must comply but also individual countries are introducing further requirements for both new and existing buildings.

Energy Performance of Buildings Directive

DIRECTIVE 2002/91/EC OF THE EUROPEAN PARLIAMENT AND OF THE COUNCIL of 16 December 2002 on the energy performance of buildings

The Energy Performance of Buildings Directive (EPBD) is the European driver for improving the energy performance of all buildings in all European countries. Each country is required to translate the Directive into national legislation according to an agreed timetable.

The EPBD is designed to provide information on the energy performance of a building to prospective buyers and tenants, or in the case of 'public' buildings, the information is to be displayed inside the building to inform all visitors. The objective is to raise awareness of the energy use of buildings, allowing prospective buyers and tenants to make informed decisions, and encouraging building owners and occupiers to improve the energy performance of new and existing buildings.

The main components of the EPBD relevant to existing buildings are given in the two boxes below.

Article 6: Existing buildings

Member States shall take the necessary measures to ensure that when buildings with a total useful floor area over 1000m^2 undergo major renovation, their energy performance is upgraded in order to meet minimum requirements in so far as this is technically, functionally and economically feasible.

Article 7: Energy performance certificate

1. Member States shall ensure that, when buildings are constructed, sold or rented out, an energy performance certificate is made available to the owner or by the owner to the prospective buyer or tenant, as the case might be.
2. Member States shall take measures to ensure that for buildings with a total useful floor area over 1000m^2 occupied by public authorities and by institutions providing public services to a large number of persons and therefore frequently visited by these persons an energy certificate, not older than 10 years, is placed in a prominent place clearly visible to the public.

Implementation: The example of UK

Article 6 of the EPBD is translated into UK legislation via Building Regulations.

Building refurbishments that accompany a 'change in use' are subject to energy use, and other provisions, of planning and Building Regulations control. Replacements to building envelope components, including windows, ventilation equipment and mechanical and electrical services, must comply with the Building Regulations 2000. Additionally, building energy performance must be improved when major refurbishments are conducted for buildings over 1000m^2. Energy requirements for building extensions and the commissioning of services are also included in Building Regulations. Large-scale refurbishments, or those that lead to a change in use of the building, may result in the project being considered equivalent to a new-build.

The energy efficiency provisions of Building Regulations will continue to be used by government to drive improvements in the existing building stock; it is the intention that forthcoming updates planned for 2010 and 2013 will further strengthen the energy performance requirements of refurbished buildings.

Article 7 of the EPBD requirements have been transposed into British legislation and are coming into force in England and Wales between 2006 and 2011. Legislation will be coming into force in Scotland and Northern Ireland over a similar period.

Energy Performance Certificates (EPCs) are based on the Asset Rating (a calculated annual energy consumption based on a standard use of the building). EPCs provide the building's relative energy efficiency in a similar form to domestic product energy ratings, and must be provided to prospective buyers or tenants when a building is constructed, sold or rented.

Display Energy Certificates (DECs) are based on the Operational Rating, that is the measured annual energy consumption of the building. DECs are required for buildings over 1000m^2 occupied by public authorities or other institutions that provide public services to a large number of people, from October 2008. DECs must be prominently displayed in the building to inform visitors. The government has committed to consult on the possible widening of these requirements to privately owned or occupied public buildings including retail outlets, cinemas and hotels. EPCs and DECs are produced by accredited energy assessors and are accompanied by a report detailing voluntary options for improving the energy efficiency of the building.

Using other legislation in the UK

There has been a clear trend for at least the last decade to reduce energy consumption in the existing building stock, with Building Regulations, planning policies and requirements regarding the provision of information to stakeholders all progressively strengthening.

Planning policies. Local planning policies supported by government legislation and Regional Planning increasingly require developments to use local heating networks (where available) and/or to use renewable energy sources (on, near or off site) to supply a proportion of their energy requirements. Exact local planning requirements regarding building refurbishments can differ significantly between local authority areas. As such, local planning requirements should always be consulted prior to any refurbishment.

Carbon Reduction Commitment. The Carbon Reduction Commitment (CRC) is a new scheme, announced in the Energy White Paper 2007, which will apply mandatory emissions trading to cut carbon emissions from large commercial and public sector organizations (including supermarkets, hotel chains, government departments, large local authority buildings), with carbon reductions of 1.1 Megatonnes of carbon per year expected by 2020. The Department for Environment, Food and Rural Affairs (DEFRA) is currently determining how the CRC will operate, with implementation expected in January 2010.

Voluntary schemes and drivers

Various voluntary standards have also been developed for reducing energy use in existing buildings; such standards are often a precursor to further mandatory controls. Voluntary schemes may be strong drivers for refurbishing buildings to low carbon standards, and are used by some organizations to impose standards on buildings they occupy. In the UK BREEAM (Building Research Establishment Environmental Assessment Method) has for many years been accepted by industry as a general standard for assessing the environmental sustainability of non-domestic buildings, has been an important driver for the improvement of the building stock and has been widely used for promotional purposes. Many organizations have environmental policies and regularly report on their Corporate Social Responsibility (CSR). Carbon emissions form a key element of this, with energy efficiency credentials often highlighted as an indicator of a responsible approach in the community. Year on year improvements in reducing carbon emissions are usually a component of this reporting, with upgrading existing building stock a common action item.

The UK Energy Efficiency Accreditation Scheme (EEAS) is the leading independent emission reduction award scheme in the UK and is open for both commercial and public sector organizations. The Scheme provides advice for improving energy efficiency and requires demonstrated improvement in energy performance to secure accreditation. By gaining and maintaining accreditation, organizations involved in the Scheme are able to raise the profile of delivered energy and carbon reductions both internally and externally, and can benefit from ongoing support to deliver further emission reductions through membership of the Accredited Organizations Network.

References

Baker, N. and Standeven, M. (1994) 'Thermal comfort in free running buildings', *Energy and Buildings*, vol 23, pp175–182

Part Two **Practice**

2 Floors

2.1 Solid ground floors

Most solid ground floors being considered for refurbishment will be non-insulated.

Original floor: Solid ground slab with screeded finish.

There is some uncertainty about the actual insulation value of non-insulated ground floors. It is very dependent upon the properties of the subsoil. The literature provides values ranging from 0.3 for large buildings to 1.0 for small shallow-plan buildings. The dependence on size is due to the three-dimensional nature of the heat flow. The outcome is that large buildings may have relatively low floor U-values already, and the cost benefit of floor insulation may be poorer than for other parts of the envelope.

Insulation options

Option 1: Load-bearing insulation above slab with reinforced screed above.

This provides some insulated thermal mass, which will offer some of the beneficial functions of thermal storage associated with heavyweight construction. The beneficial effects of thermal mass will be fully realized if dense conductive materials (e.g. ceramic tiles) are used as a floor finish, but reduced if finishes such as carpet are used. For screed thickness of up to 75mm, this amount of thermal storage would be significant for 24-hour cycles only, due to its isolation from the thicker ground slab.

Option 2: Load-bearing insulation above slab with lightweight decking above.

This behaves as a lightweight construction since the mass is isolated by the insulation. The floor finish will have little effect on thermal response.

Option 3: Raised floor with rigid or non-rigid insulation (quilt) on original floor.

Raised floors are used where access to communications wiring and services are required across the whole floor. They may also be of value where underfloor voids are to be used as part of a natural ventilation system. It must be noted however, that with wireless IT technology the demand for raised floors for IT servicing has diminished.

Option 4: Replaced slab with rigid insulation beneath.

This would only take place in major refurbishment, or in new parts of a building. It offers both high insulation and large thermal mass.

These options are illustrated in Figure 2.1.

Underfloor heating or cooling

Underfloor heating (or cooling) pipes can be incorporated in floor options 1, 2 and 4. The thermal mass of the screed in option 1 will result in a slow response emitter which could lead to control problems where rapid changes in heat loads and gains are expected. Option 4 will have a very slow response (days rather than hours), and would give control problems in all but continuously occupied buildings with very constant gains profiles.

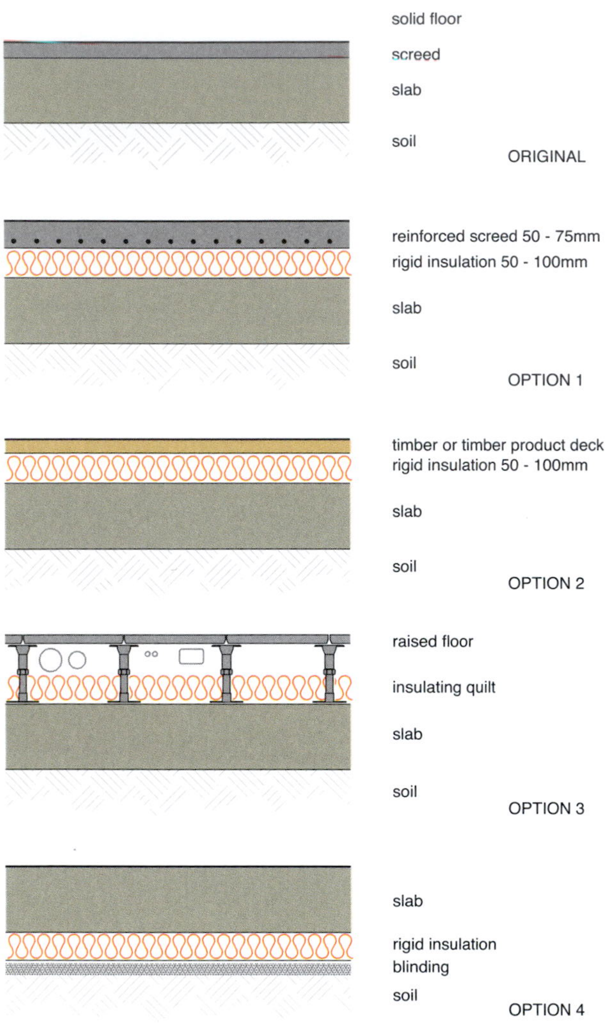

Figure 2.1 Insulation options for solid ground floors

For option 2, the heating pipes are located just beneath the decking in the surface of the rigid insulation. This results in a rapid response emitter with a large surface area, which with suitable controls can be very efficient.

Underfloor heating with option 3 could be achieved with a warm-air supply.

Underfloor heating should never be installed without insulation from the ground.

2.2 Suspended ground floors

The insulation value of the non-insulated suspended floor is dependent on the degree of ventilation of the underfloor void. For traditional timber floors, this is often quite high and results in the U-value being significantly higher – typically around 1.5 – than for a solid floor. Reducing the ventilation rate would reduce this but would lead to high humidity and subsequent decay of the timber. For non-timber floors the void is also normally ventilated, to avoid condensation.

Original floor A – screed on hollow ceramic pots between steel joists, or concrete blocks between reinforced concrete joists, over void or crawl space

Original floor B – timber (solid or timber-derived product) decking on timber joists

Insulation options

For floor A as options 1, 2 and 3 for solid floors, with similar thermal behaviour, and (Figure 2.2):

Option 5: If access to crawl space permits, it may be possible to apply insulation to the underside of the floor, or lay it onto the ground or oversite concrete.

For floor B – insulation, or; as option 2

Option 6: Remove deck and apply rigid or semi-rigid insulation between joists.

Cold bridging is tolerable due to relatively high thermal resistance of timber. Thermal behaviour as lightweight.

Underfloor heating or cooling

Heating pipes can be incorporated in the above deck screed or insulation as in options 1 and 2. Pipes can also be installed in the top surface of the between-joist insulation in option 5.

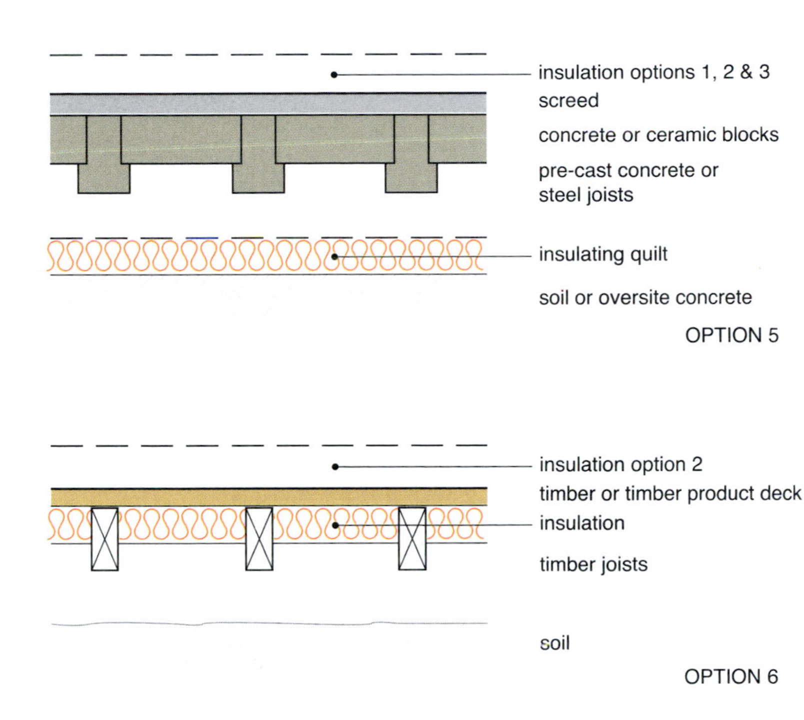

Figure 2.2 Insulation options for suspended ground floors

Apart from the opportunities offered for heating and cooling systems integrated with the floor structure, insulating the floor can have a considerable impact on the dynamic thermal response of the building.

The simple rule applies, that thermal mass in the floor only affects the interior if it is coupled to it. This means that options that isolate the floor surface from the massive structural slab (as in options 2 and 3) will change the thermal response of the space, making it much quicker to respond to heat inputs.

This may be advantageous in responding quickly to heating systems in intermittently occupied spaces, or disadvantageous in spaces that might receive large and irregular heat gains, such as solar gains. The latter has implications for comfort, where a traditional paved solid floor maintains a steady temperature well below peak daytime temperatures, encouraging both radiative heat loss and temperature stratification, both welcome in warm conditions. Massive floors are also effective at absorbing solar gains when the sunpatch is located on the floor, thereby reducing daytime overheating and increasing the usefulness of the gains to reduce auxiliary heating later in the day.

Option 1, although isolated from the original structural slab, provides a significant amount of thermal mass, and this will behave similarly to the original non-insulated slab to fluctuations of a diurnal frequency, but will not show the long-term stability associated with thick slabs in ground contact.

2.3 Intermediate floors

Although heat flows may occur due to temperature differences between floors, these will be of far less significance than heat losses from the building envelope. Thus, it will not be cost effective to insulate intermediate floors, except in special circumstances, where it is known that significantly different temperatures will be maintained. Options would be similar to suspended ground floors.

2.4 Thermal response implications of floor insulation

Floor insulation reduces heat losses in the heating season and may also improve comfort by reducing temperature stratification where cool air collects close to floor level. The impact of floor insulation on heat loss may be less than expected when applied to deep floor plans, due to the relatively low effective U-value of the non-insulated floor away from the perimeter.

Walls

Wall insulation is important for:

- heat retention in cool conditions;
- heat exclusion in warm conditions;
- preventing the ingress of solar gains made by the absorption of radiation on the outside of the opaque wall.

In non-extreme climates, due to the seasonal variation, all of these conditions will be encountered during the year. This means that in general, wall insulation will always bring positive benefits. There may be some situations, however, where benefits may small in comparison with the cost, or present technical difficulty and/or unacceptable visual impact. This is sometimes the case in historic buildings. Or there may be buildings where due to high internal gains, the lack of need of close temperature control or the nature of the activity, heating and cooling loads are already modest.

There may also be some circumstances where overheating is made worse. This can occur in buildings with high internal gains in cool climates, particularly in lightweight buildings, or if the insulation is positioned on the inside of envelope. However, except in rare occasions where consistently high internal gains are present – for example, due to a manufacturing process – insulation will provide a better annual performance, and overheating can be addressed by proper ventilation and gains control. This is demonstrated in section 1.6, p8.

3.1 Solid walls

Here we refer to solid walls constructed of bricks, stone, concrete block or in situ concrete.

Other materials that might be encountered in historic buildings could include materials such as rammed earth (cob or adobe), and timber framed walls with solid infill of mud, clay, soil composites, often reinforced with light timber sections (e.g. wattle and daub). These kinds of constructions often present problems for refurbishment due to damp ingress and decay. Solutions are highly specific to the technical details of the construction, and are not dealt with here.

The U-value of the non-insulated wall is of course dependent on the material and thickness of construction. Typical values range from 1.0 to 3.0, which fall a long way short of current new-build values.

External insulation (Figure 3.1)

Option 1: Rigid insulation material fixed to wall and render applied.

Option 2: Framing fixed to wall to create voids for non-structural insulation, render applied on support layer (e.g. metal lath).

Option 3: As above but with rigid cladding applied (e.g. timber, metal panel).

Option 4: Composite engineered cladding panel providing weathering, insulation and structural support.

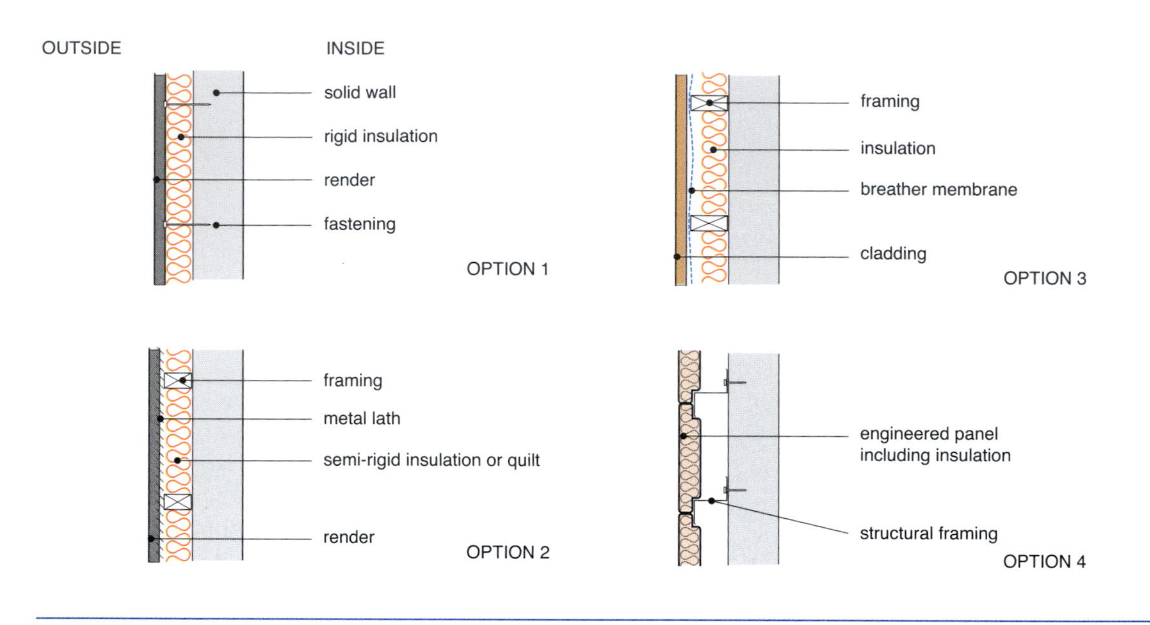

Figure 3.1 Options for external insulation of solid walls

Implications for external insulation

All forms of external insulation can be applied without changing the thermal response of the interior. This is because the thermal mass of the structure remains coupled to the interior, which is where the gains are made (solar gains through windows or gains from internal equipment and occupants) (Figure 3.2).

External insulation also protects the structure from solar gains made on the external surface of the building. These are important considerations as both can reduce the need for air-conditioning. Finally, external insulation may be part of a treatment to provide new weatherproofing to a degraded wall.

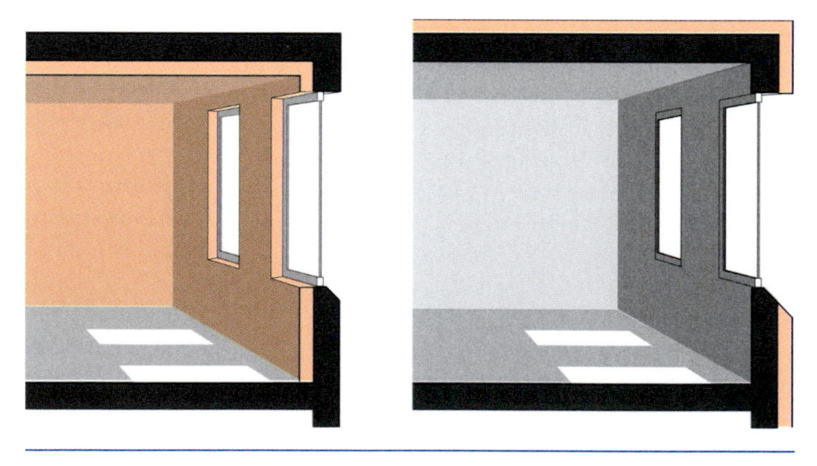

Figure 3.2 Coupling of thermal mass with room interior

In most cases, externally applied insulation eliminates cold bridges and (unlike internal insulation) does not create new cold bridges. The exception to this is where a balcony or other structure will protrude through the insulating layer. This may not be easy to solve without reconstructing the attachment with a high strength, low cross section element.

Non-thermal advantages include the benefit of work being able to be carried out without disturbing the interior, and possibly allowing occupation to continue.

External cladding may cause a major visual impact. For some historic buildings this will be unacceptable. In cases where buildings are already of rendered finish, using options 1 or 2 could leave the building with no significant change of appearance. However, applying external insulation to facades that are articulated and have openings for windows etc., will be technically challenging.

In other cases, change of appearance may be welcome, and options 3 and 4 are often used to give visual as well as thermal improvement.

Internal insulation (Figure 3.3)

The constructional options for internal insulation are similar in principle to external insulation.

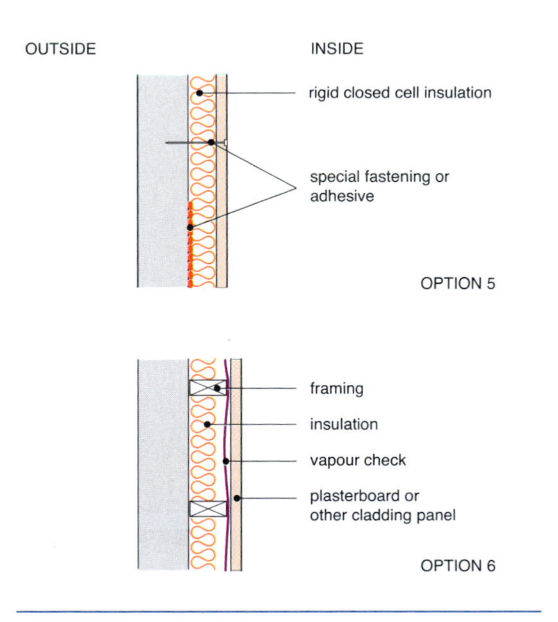

Figure 3.3 Options for internal insulation of solid walls

Option 5: Rigid insulation material fixed to wall and render or plasterboard applied. Plasterboard with integral insulation is available.

Option 6: Framing fixed to wall to create voids for non-structural insulation, plasterboard or other cladding panel.

However, option 4 will rarely find application – one fundamental difference being that the final internal finish does not have to be a weathering finish.

Thermal response

The *thermal response* of the interior is changed since the thermal mass of the original structure is now isolated from the interior (see Figure 3.2). This will have most impact in smaller buildings where a greater proportion of the walls of the rooms is external wall. Where a large building is highly subdivided by partitions, if these are heavyweight construction, and if further mass is present and accessible in the floor and ceiling, the effect of insulating the external wall will be slight. It is difficult to make quantitative rules, and these issues should be investigated by thermal simulation.

Cold bridges

Cold bridges will be created where internal partitions and floors meet the external wall (Figure 3.4). If, as is usual, they are of conductive materials, such as masonry or concrete, this is a difficult problem to overcome. The solution necessitates bringing the insulation back for a distance from the external wall or floor (Figure 3.5) which in itself causes problems particularly in the case of the floor.

Cold bridges lead to increased heat loss, since the average U-value of the envelope is increased. However, in walls, this effect may be relatively slight, and could in principle be compensated by extra insulation. The greater importance of cold bridges is that they lead to lower surface temperatures and may become sites for surface condensation. This is particularly prevalent in

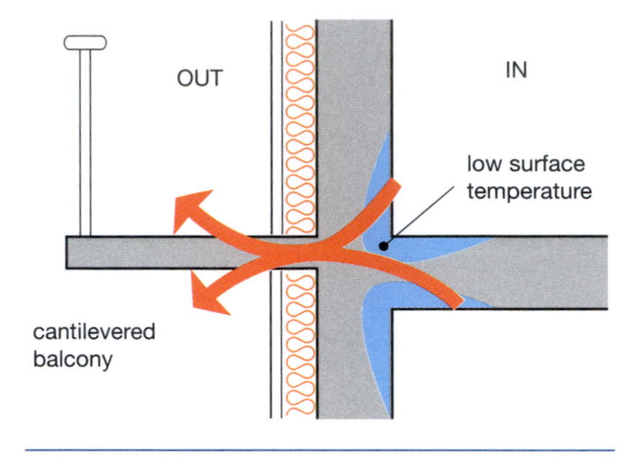

Figure 3.4 Cold bridges left by external insulation

housing, or in other building uses with high occupancy and moisture production.

However, not all cold bridges have equal thermal impact, and in some circumstances may be acceptable. The actual heat loss and resulting depression of surface temperature can be calculated from the geometry and material properties of the bridge, using three-dimensional heat flow analysis. This could be worthwhile in a large project where the elimination of cold bridges could be very expensive.

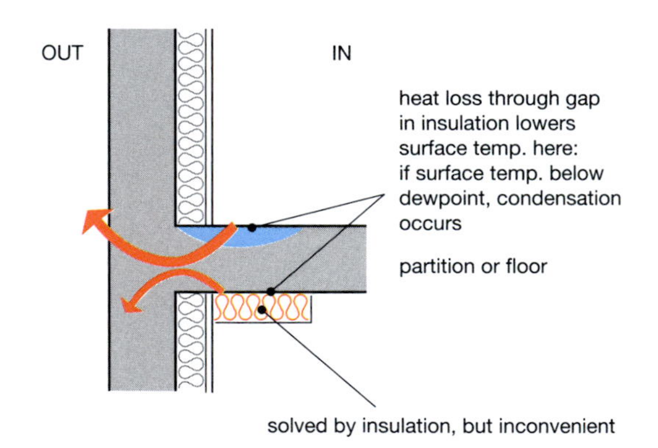

Figure 3.5 Insulating a cold bridge will be costly and inconvenient

Interstitial condensation

Internally applied insulation has the potential for creating *interstitial condensation* – that is, condensation occurring inside the structure. This can have a very damaging effect on the structure, causing corrosion and decay, and in some cases reducing the effectiveness of the insulation.

The cause of interstitial condensation is illustrated in Figure 3.6. It shows that the diffusion of water vapour through the structure to a part which is at a temperature below the dewpoint[1] of the air. The solution is simple in principle – prevent the vapour from diffusing through the material by applying a vapour check barrier to the warm side of the insulation.

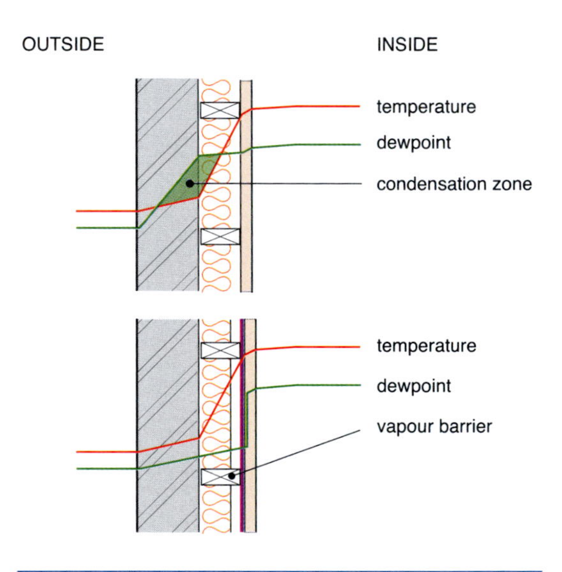

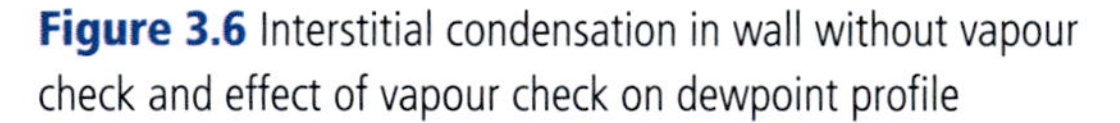

Figure 3.6 Interstitial condensation in wall without vapour check and effect of vapour check on dewpoint profile

In practice it is not simple and two difficulties arise. Firstly it is difficult to ensure that the vapour check is unperforated and sealed to such as door linings. Secondly, pressure differences

1 As a given mass of air is cooled down it can hold less and less water as vapour. The dewpoint is the temperature at which it has a relative humidity of 100 per cent and further reduction of temperature will result in condensation.

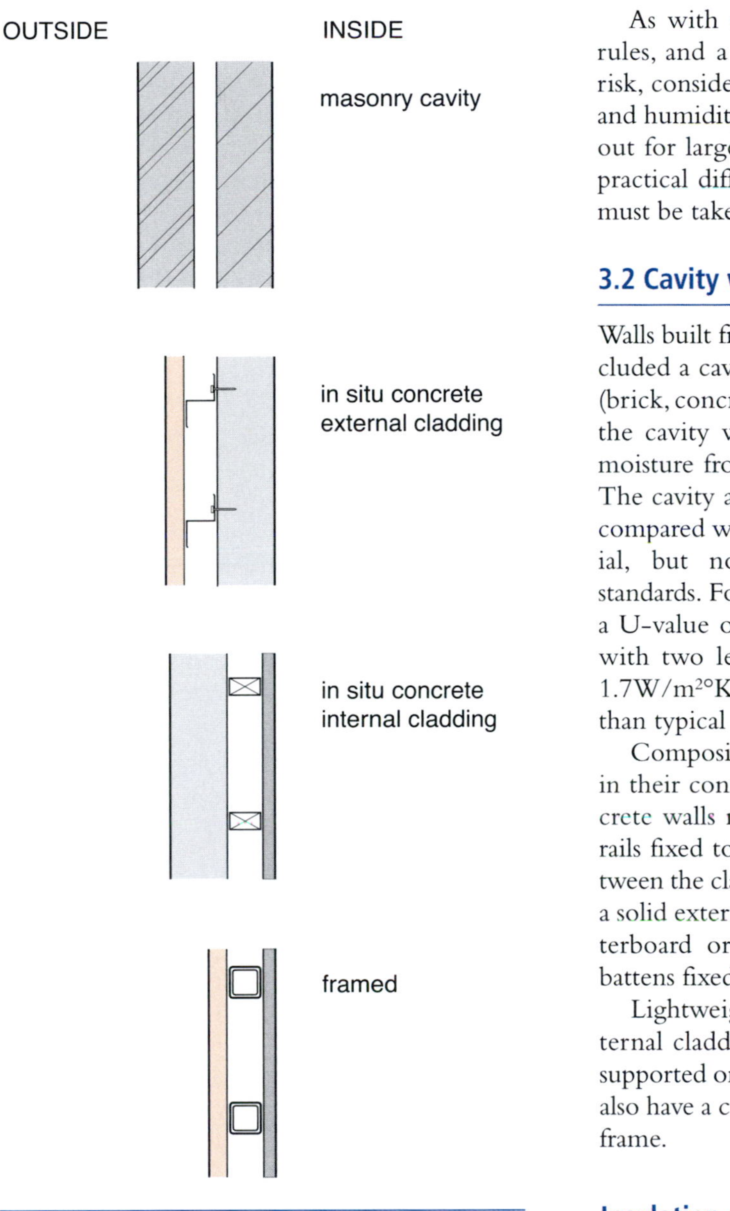

Figure 3.7 Typical cavity wall constructions

across the building envelope (due to wind or temperature differences) can cause bulk flow (non-diffusive) of air via voids, conducting moist air to cold sites. This is likely to be of much greater magnitude than diffusive flow.

As with cold bridges, there are not simple rules, and a dynamic analysis of condensation risk, considering the evolutions of temperature and humidity in the structure, should be carried out for larger projects. In making this analysis, practical difficulties of installing vapour checks must be taken into account.

3.2 Cavity walls

Walls built from about 1950 onwards usually included a cavity. In walls of double leaf masonry (brick, concrete, stone, etc.) the main purpose of the cavity was to prevent the transmission of moisture from the outer leaf to the inner leaf. The cavity also increased the thermal resistance compared with the same amount of solid material, but not sufficiently to meet modern standards. For example, a 225mm solid wall has a U-value of $2.3W/m^2{}^\circ K$, whilst a cavity wall with two leaves of 112mm has a U-value of $1.7W/m^2{}^\circ K$. This is still at least five times higher than typical newbuild values.

Composite walls may have cavities inherent in their construction. For example, in situ concrete walls may have a cladding hung on steel rails fixed to the concrete, creating a cavity between the cladding and structural wall. Walls with a solid external leaf may be 'dry lined' with plasterboard or timber composite supported on battens fixed to the structural wall.

Lightweight walls that are composed of external cladding panels and internal 'dry lining' supported on steel, concrete or timber frame will also have a cavity, usually enclosing the structural frame.

Insulation options

The insulation value of the cavity is dependent on the ventilation and air movement. If the cavity is freely ventilated to the outside – for example, as in 'rain-screen cladding' – the insulation value of the cavity is virtually zero. The more restricted the ventilation, the greater the insulation value, as the temperature in the cavity

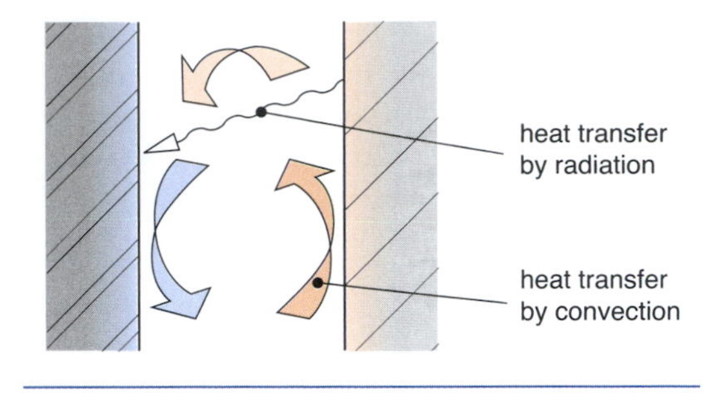

Figure 3.8 Heat transfer across a wall cavity by convection and long-wave radiation

increases. However, even for a non-ventilated cavity, the insulation value is limited by the transfer by convection and radiation of heat from one leaf to the other, as in Figure 3.8.

Improvement in insulation value can be made in two ways:

1. Reducing the radiative transfer by including low-emissive surfaces.
2. Reducing or eliminating convective transfer by filling the cavity with insulating material.

Note that these are alternatives; there is no advantage using low-e surfaces if the cavity is full, since heat transfer is now entirely by conduction.

Practical considerations

Masonry walls

In the case of double masonry walls, it is usually impractical to remove one leaf in order to place rigid insulation in the cavity. Thus the cavity can only be considered as a location for insulation if material can be injected or blown into the cavity via small openings. This is now a common practice, particularly in domestic buildings, using materials such as rockwool, glass fibre and expanded polystyrene beads. Other materials include recycled cellulose fibre and vermiculite.

With this technique, the cavity becomes filled, and there is concern that the moisture-isolating function of the cavity will be lost. The nature of the climate, the degree of exposure of the building to driven rain, and the condition of the outer leaf all have to be considered for the particular project, and specialist knowledge sought.

It is relevant to point out that in newbuild masonry cavity construction, it is normal practice to fix semi-rigid insulation to the inner leaf and leave a cavity of at least 25mm between the insulation and the outer leaf.

Composite and lightweight walls

It is less common practice to inject walls of this type with loose fill, although if the outer leaf is reliably waterproof, and the conditions relating to interstitial condensation (see below) are satisfied, then there is no reason, in principle, why it should not be done.

The refurbishment of this wall type may involve the stripping of either the inner or outer leaf. Provided the cavity is large enough this will give the opportunity to fix semi-rigid insulation, still maintaining a cavity.

Interstitial condensation

The principle of interstitial condensation control is to keep the temperature gradient (from inside to outside) above the dewpoint gradient. This can be achieved by vapour check membranes as already discussed, and illustrated in Figure 3.6.

Clearly the incorporation of a vapour check membrane is not possible in existing masonry cavities. When a cavity is insulated, the inner leaf is usually sufficiently impermeable to water vapour to ensure that the temperature in the insulation does not drop below the dewpoint, as in Figure 3.9. Note that the drop of dewpoint in an element is dependent on its relative resistance to vapour compared with the other layers. However, there could be combinations of a very permeable inner leaf with an impermeable outer leaf – for example dense concrete or granite – where condensation could occur on the inside of the outer leaf. This would be a particular concern if a biodegradable material such as reclaimed cellulose fibre were to be used. The situation could be improved by rendering the inner leaf more impermeable to vapour by surface treatment such as paint.

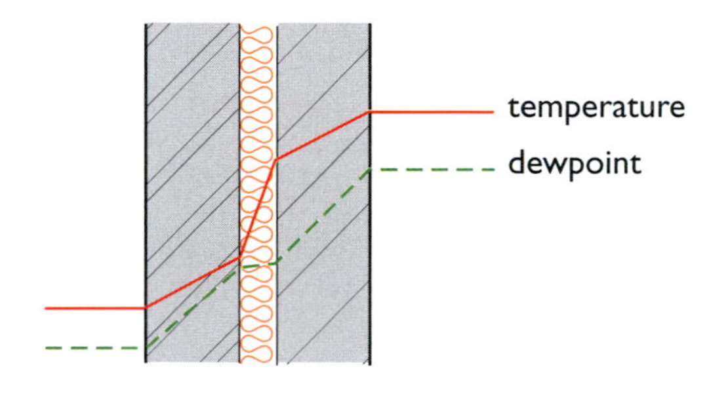

Figure 3.9 Temperature and dewpoint profiles in masonry wall with insulated cavity

In composite and lightweight walls, the situation is similar to the case described for internal insulation of solid walls. The risk of interstitial condensation will often rule out the retro filling of the cavity unless the inner leaf is known to have a much higher vapour resistance than the outer leaf. This is definitely not the case where the inner leaf is a dry lining such as gypsum plasterboard, or where internal panelling has joints through which there could be bulk air flow. In these cases, the only solution for cavity insulation would be to strip off the inner cladding and install a vapour check, or replace the cladding with an inherently vapour resistive material such as polyurethane foam.

Where the problems associated with filling the cavity cannot be overcome, external or internal insulation could be applied. The thermal implications of this action would then become the same as the case of the solid wall.

Thermal implications

The principles of the change in thermal response when applying insulation are the same as described for solid walls. If existing mass is isolated from the room, by the insulation, then thermal response time will be reduced according to the proportion of room surface treated.

In a double leaf masonry construction, most of the active thermal mass is in the inner leaf and so filling the cavity will have very little effect on the thermal response.

The only point to add is that if insulation is added to a lightweight wall, there will be no change to the thermal response other than reduced heat loss.

Retrofit inner or outer leaf

During refurbishment of ancient or historic buildings there may be the need to build an inner leaf, on its own foundations, in order to provide structural support to upper floors, or to the existing external wall. If a cavity is to be included, then, in principle, this could be insulated. However, it is possible that the requirements for bonding the new leaf to the existing wall may require the cavity to be too small for significant insulation to be included. In these cases, other insulation options for solid walls will have to be adopted.

4 Roofs

Roofs include a wide range of construction types and thus refurbishment solutions tend to be very specific. Clearly a vital function of roofs is the exclusion of water, and refurbishment is often driven by this. When this leads to re-roofing, that is, replacing the whole weathering element, several options for thermal improvement will present themselves.

In other situations, the weathering function of the roof may be perfectly satisfactory, and the intervention may be to improve the thermal performance only. In this instance a more careful appraisal of benefit will have to be made and cost effective options may be more limited.

Roof insulation is important for two main reasons:

1. A poorly insulated roof can be a source of large heat losses due to its exposure to wind, the large convective transfer from a warm surface upwards, and high radiant losses to the night sky.
2. The roof surface receives the greatest insolation during the summer period – solar gains conducted through the roof can be a major cause of overheating and possibly expensive mitigation by air-conditioning.

However, it has to be borne in mind that improvements to the roof will only have significant benefit to the floor below. Thus the overall impact of roof improvements will be quite different for, say, a single-storey building and an eight-storey building.

Roof types

The categories below have been chosen to relate to opportunities for insulation:

1. roofs with accessible attic spaces (double pitched, mono-pitched or flat);
2. roofs with voids;
3. solid roofs.

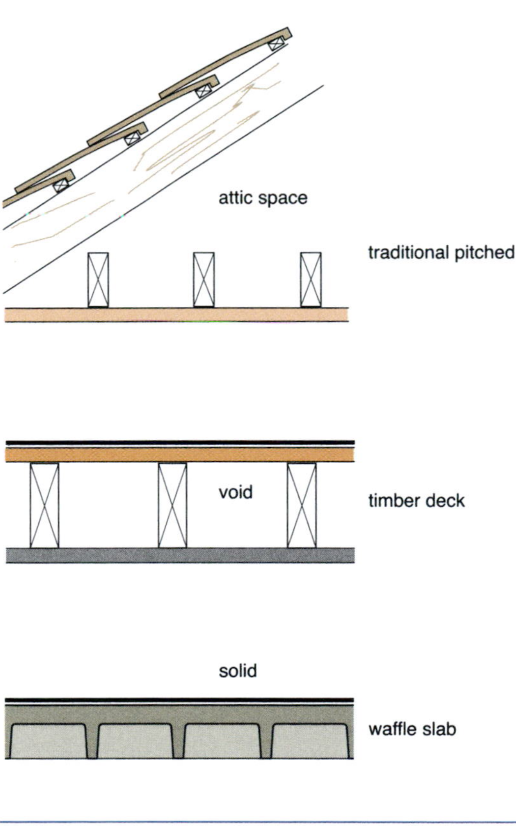

Figure 4.1 Types of roof

4.1 Insulating roofs with attic spaces

These give access to the upper surface of the ceiling element, thereby allowing the placement of non-rigid insulation material. Usually there is no space limitation, thus large thicknesses of insulation can be accommodated and high standards of insulation achieved at low cost. For example, the application of 300mm of fibreglass to an uninsulated plaster ceiling under a tiled roof will reduce the U-value from 2.5 $W/m^{2}{}^{\circ}K$ to 0.1 $W/m^{2}{}^{\circ}K$, a factor of 25 times.

Ventilation of attic space

In traditional construction it is likely that the ceiling will be lime or gypsum based plaster supported on lath or plasterboard. These are vapour permeable. Many modern ceiling constructions are also permeable, due to gaps and joints, even if the material itself is not vapour permeable. Since water vapour is generated in all occupied buildings, it is essential that the attic space, into which the water vapour will migrate by either diffusion or bulk flow, must be ventilated sufficiently to keep the dewpoint below the air temperature and surface temperature.

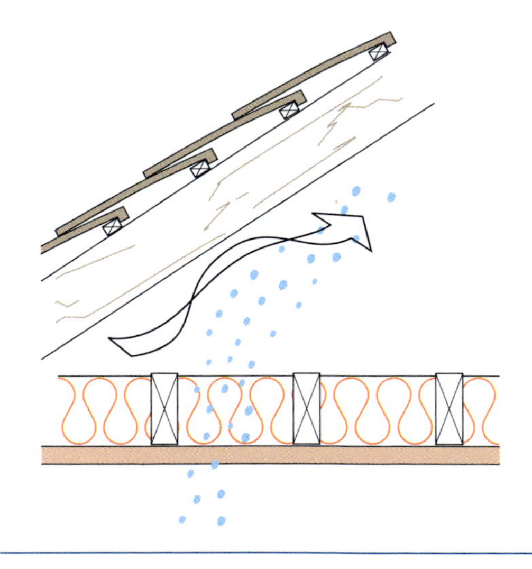

Figure 4.2 Migration of water vapour through ceiling and insulation requires ventilation of attic space

In practice, this may not be possible at all times because the temperature of the external roof element may drop below air temperature due to radiant losses to the sky, particularly at night. However this will be a transient effect, and any condensation will evaporate at a warmer part of the diurnal cycle, and be removed from the attic space if there is sufficient ventilation.

It is difficult to give simple rules on area of ventilation openings since there are unknown variables such as the humidity in the building and the vapour permeability of the ceiling. In some cases, attic ventilation provision is governed by building codes. However, as an approximate guide the total ventilation area should be between 0.3 and 1.0 per cent of the ceiling area, well distributed in plan and vertically. Distribution is important to encourage good mixing and stack effect ventilation on calm days.

Care should be taken to ensure that the insulation material does not obstruct air flow at the eaves, particularly when there are no other ventilation openings. As a general rule, it is much safer to over-ventilate, since as the attic space is cold (all of the temperature drop is in the insulation), there is no energy penalty for over-ventilation.

4.2 Insulating roofs with voids

This will commonly apply to flat roofs. It will also apply to pitched roof attics that are to be occupied, where the insulation must be in the plane of the outer roof element.

If the void is to be used as the location for insulation, the main problem is interstitial condensation, as described for walls in section 3.1, p34. The problem is exacerbated for flat roofs, since the horizontal outer element (typically bitumen, lead or a rubber/plastic based membrane) has to be inherently waterproof and thus highly resistant to vapour. This rules out the filling of a roof void with particulate insulation (e.g. mineral fibre, polystyrene beads) by injection through small openings, since it would be impossible to insert an internal vapour check between the ceiling and insulation. Furthermore, unless the void is unusually large and only partially filled, sufficient ventilation will be very difficult to provide passively.

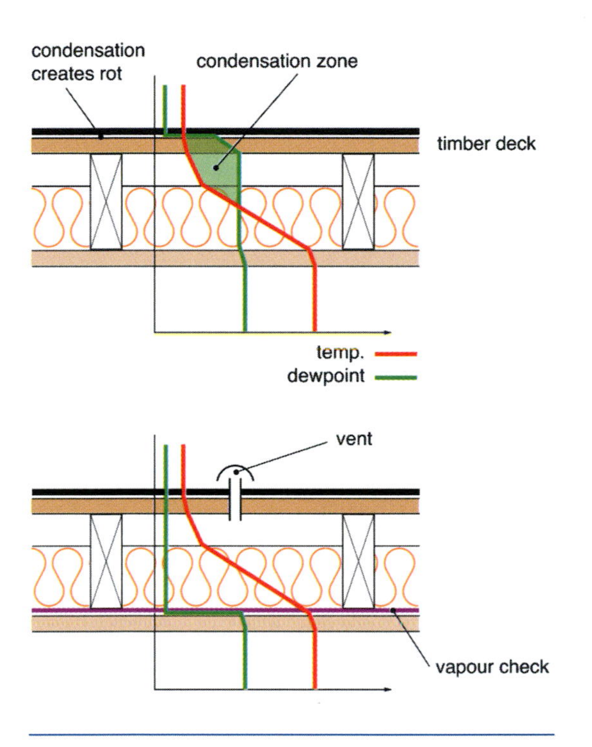

Figure 4.3 Dewpoint profiles in insulated flat roof with and without vapour check and roof ventilator

Where the inner and outer roof elements are pitched – for example, a tiled roof over an attic room – the outer element has only to be water-shedding rather than water-proof. Provided a vapour-permeable membrane (or no membrane at all) has been used between the outer layer and the void, it might be possible to inject loose fill insulation without a vapour barrier. The risk of condensation will depend, as with 'attic spaces', above, on the prevailing humidity in the heated space, and the vapour permeability of the inner ceiling. However, the risk is high and it is not recommended without careful analysis.

In both the above cases, the recommendation is to gain complete access to the void by the removal of either the inner or outer elements, and install a vapour check membrane on the warm side of the insulation. The technical details of installation are not dealt with here, but it is essential that it is carried out with a high standard of workmanship and supervision.

Insulation materials can be highly vapour permeable – such as mineral wool or cellulose fibre – or inherently impermeable, such as closed cell polystyrene or urethane foam. Closed cell foams will not suffer interstitial condensation within their volume. However bulk flow of air can take place between slabs of insulation and conduct moist air to cold parts of the structure – thus a vapour check is still essential. It is also strongly recommended that the void is not filled, leaving a space above the insulation that can be ventilated.

There can be no short cuts to this problem. There have been cases of total structural failure due to interstitial condensation.

4.3 Insulating solid roofs (or roofs with inaccessible voids)

For flat roofs three options are available:

1. insulation above the waterproof membrane;
2. insulation between waterproof membrane and structural deck;
3. insulation below the structural deck.

Insulation above the waterproof membrane

The so-called 'upside-down' roof relies on the inherent waterproof properties of closed-cell plastic or glass foam insulation. Slabs of rigid insulation are cut to fit and simply laid on the existing waterproof membrane. They are usually weighted down with either concrete slabs or gravel. This method is particularly suited to concrete decks, where the extra loading due to the gravel or slabs is unlikely to present problems.

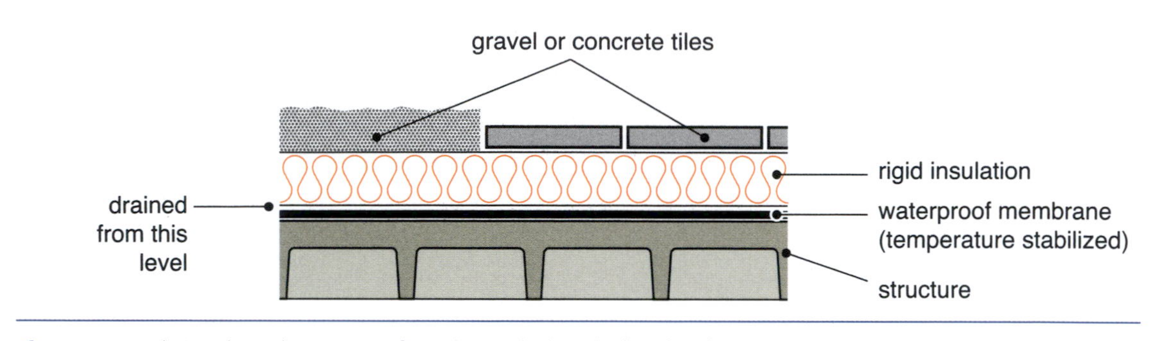

Figure 4.4 Insulation above the waterproof membrane; the 'upside-down' roof

The roof is drained at the membrane level.

The great advantage of this method is that the waterproof membrane is protected from thermal stress and other damage, and kept at a very stable temperature. The work can also be carried out with no disturbance to the interior, and because the waterproof membrane also acts as a vapour check, and is on the warm side of the insulation, there is zero risk of interstitial condensation.

A further advantage is that most cold bridges (see section 4.4) will be avoided, although some problems may be encountered where there are parapet walls.

Insulation between waterproof membrane and structural deck

This might be carried out on a roof that had to be stripped due to the poor condition of the waterproof membrane.

The disadvantage of the method is that the waterproof membrane remains exposed to the weather. In the case of heavy concrete or masonry based roofs, the waterproof membrane will experience much greater temperatures and temperature gradients due to the isolation of the membrane from the thermal mass of the structure. This can develop shear stresses along shadow lines and at the edges of puddles that, together with the effects of UV, can accelerate the ageing and failure of the membrane. On roofs with lightweight decks, the difference will be less marked, although even the thermal capacity of timber will have some moderating effect.

Thus wherever possible, it is recommended that option 1 is adopted. The only exception to this might be where on a historic building part of a flat or low-pitched roof is covered with lead or copper, and a change of appearance is unacceptable.

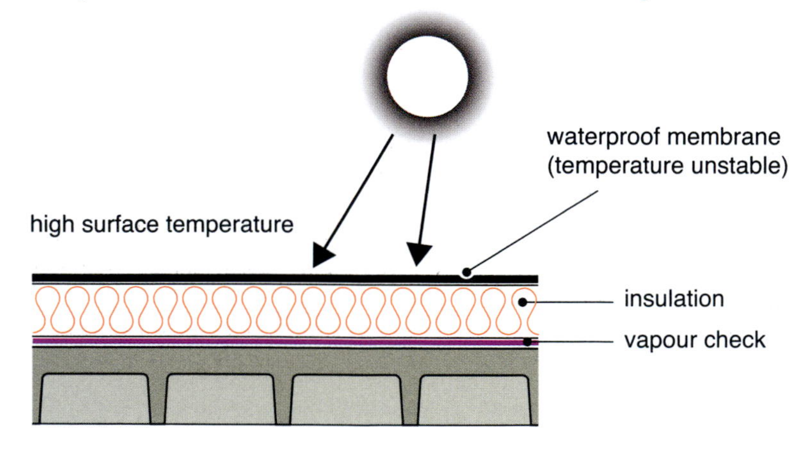

Figure 4.5 Insulation between deck and waterproof membrane leading to high thermal stresses

Insulation below the structural deck

This will almost always involve the erection of a secondary supporting structure (i.e. ceiling), and thus the issues become the same as for existing roofs with voids. The main concern is the avoidance of interstitial condensation by preventing moist air from the interior getting to the cold underside of the structural deck. This can be achieved (but not without difficulty) by using a vapour-check membrane and ventilation as described in section 4.2.

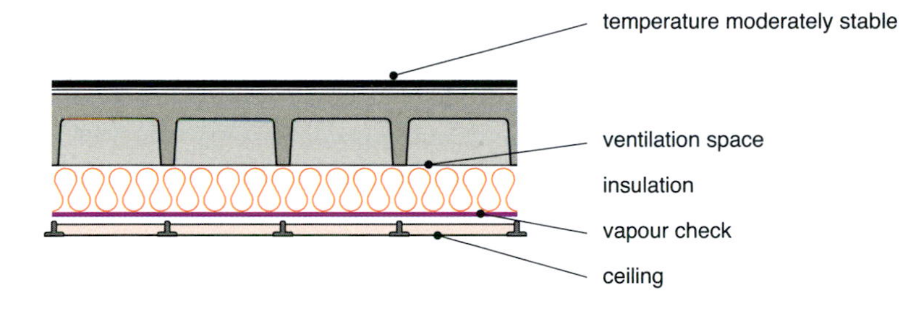

Figure 4.6 Internally insulated solid roof

An exception to this is where an inherently vapour proof material is fixed to the underside of the deck. This is sometimes achieved by spraying with an expanding polyurethane foam. Lightweight renders with cellulose fibres and other mineral fibres are also used. However, due to the vapour permeability of the render, the possibility of interstitial condensation must be much higher.

This method has also been used on lightweight pitched roofs made of a single element such as corrugated iron (or other metal), or even tiles. It is often regarded as a short-term solution, particularly in the latter case. The implications for materials such as polyurethane for the installer's health and the environment generally should be considered.

4.4 Other thermal issues

Surface reflectance

In warm climates where solar gains through the opaque fabric of the roof are a problem, the thermal performance of a roof can be greatly influenced by increasing the reflectance of the roof surface. It is common practice in tropical areas, where for traditional buildings the roof U-value is usually quite high.

The penetration of solar gains made on the outer surface of a roof construction is described by the Solar Heat Gain Factor (SHF). Table 4.1 shows the effect of changes of reflectance by simple low cost means.

Table 4.1 Solar Heat Gain Factors (SHF) for various constructions and finishes

Roof construction	Solar heat gain factor %	U-value W/m²°K
(1) Non-insulated dark coloured steel corrugated sheet with ceiling	7.6	1.9
(2) as (1) with 50mm fibreglass	5.2	1.4
(3) as (1) whitewashed ext.	3.7	1.9
(4) 150mm concrete slab	9.1	3.3
(5) as (4) with 50mm fibreglass inside	3.1	0.6
(5) as (4) whitewashed ext.	4.1	3.3

Source: Koenigsberger, O. and Lynn, R. (1965) *Roofs in the Warm Humid Tropics*, Architectural Association, London

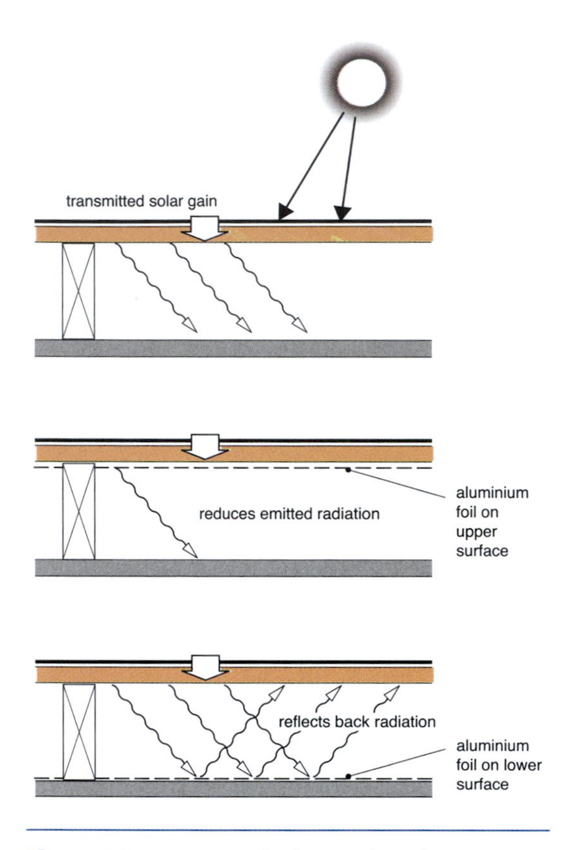

Figure 4.7 Transmission of solar gain through non-insulated horizontal roof void and effect of low-emissivity surface (aluminium foil) on upper and lower surface of the cavity

Most paints that are optically white – that is, they reflect nearly all visible radiation – still absorb the near infrared (IR), which forms nearly half of the energy in the solar spectrum. New paints that are reflective to the near IR are becoming available, and even when coloured (i.e. absorbing some part of the visible spectrum) their performance is still better than conventional white paint.

Polished metallic surfaces also have low IR absorption. However, they have poor emissivity in the long-wave IR, thus reducing the loss of heat at night. Conventional non-metallic paints have good long-wave emissivity.

When not dictated by the materials and in cases where overheating is the main problem, the use of high reflectance surface treatment is always recommended. Not only does it reduce transmitted solar gains, but also the surface suffers less from degradation due to high surface temperatures. Finally, in urban environments, the higher the urban reflectance (albedo), the less the urban heat island effect.

Low-emissivity membranes in cavities

Figure 4.7 shows the effect of improving the performance of a roof by reducing the transmission by radiation across the roof cavity or attic space by the use of low-emissivity membranes such as aluminium foil or aluminium-coated plastic. This could be a low cost measure where the roof space is accessible, but becomes increasingly less effective when the roof is insulated conductively.

It is important to note that reflective membranes are only effective in reducing transmission by radiation, and therefore only have any beneficial effect when facing a cavity. They should not be used by laying them on top of insulation unless they are of the breather variety. Impervious membranes would be likely to cause interstitial condensation.

Thermal mass

The issues are much the same as with wall insulation – that is, the effect of internal insulation is to isolate the thermal mass from the space and therefore remove the generally beneficial effect of reducing peak temperatures.

This makes the externally placed roof insulation (either above or below the waterproof membrane) doubly beneficial since it not only provides thermal mass to the interior to absorb daytime gains, but also protects the structure from daytime solar gains that would otherwise be transmitted into the interior in the evening or night.

Cold bridges

Cold bridges are areas where structural (or other components of non-insulating material) prevent the application of insulation. Typically these will be beams or joists, where insulation is placed in the space between them. The effects and treatment of cold bridges has been dealt with in some detail in Chapter 3, 'Walls', and the principles are much the same.

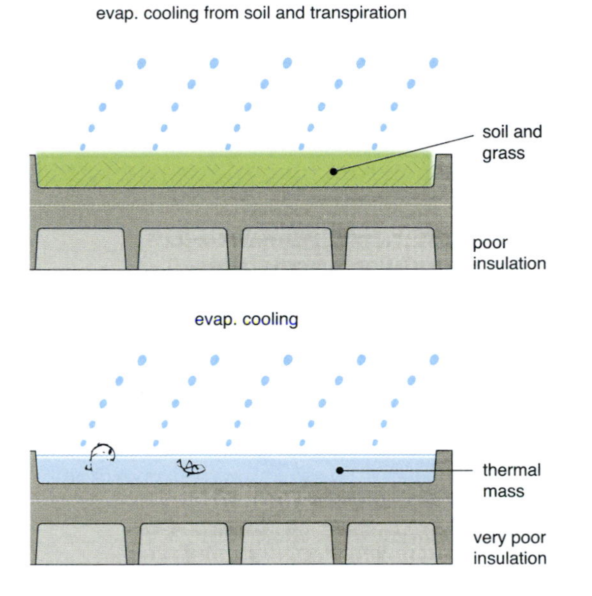

Figure 4.8 Green roofs and roof ponds add thermal mass and evaporative cooling to the roof but do not in themselves provide significant insulation

4.5 Green roofs and roof ponds

Green roofs

Green roofs consist of a layer of soil or turf which supports growing vegetation. Flat or low pitched roofs are the commonest in new building, although steeper pitched roofs can be engineered, and are actually more common in vernacular examples.

Contrary to what is often claimed, turf (soil and vegetation) is not good thermal insulation compared to modern insulation materials, as illustrated in Table 4.2. If it is wet – that is, in a condition to support growth – it is very poor. The only thermal benefit a turf roof could bring is in the provision of thermal mass. However, for buildings that are heated, thermal mass without insulation is not beneficial over the year, and once insulation is added (obviously it has to be on the inside), the benefit of the thermal mass is largely lost.

In a warm climate, where the mean temperature is somewhere near comfort temperature, non-insulated thermal mass may be beneficial, especially in the roof, which receives the largest

Table 4.2 Properties of materials for green roofs compared with conventional materials

Material	Conductivity W/m°K	Density kg/m³	Load (W/m²°K)kg/m² for U = 1.0
wet soil	1.2	1900	2280
dry soil	0.55	1500	825
dense conc	1.6	2200	352
lightweight conc	0.4	1200	480
wood	0.125	610	76
rockwool	0.036	50	1.8

solar impact, although it is likely that insulation would still be the better option. A further benefit may be gained by evaporative cooling, provided there is a source of water for replenishment.

If being considered for retrofit, the major non-thermal consideration is whether the structure can take the extra load. This will rule out the option for many lightweight constructions, which unfortunately, stand to gain most. On the positive side, the green roof can be seen as a replacement for the original vegetation that the building replaced, and in urban areas could provide a valued recreational area, as well as contributing to CO_2 absorption.

However, one important difference is that natural vegetation has access to ground water, whereas green roofs will have to have pumped water supplies due to the intermittency of supply from rain, and limited storage capacity of the turf.

Roof ponds

The roof pond is thermally one step further than the green roof, providing a larger and more responsive thermal mass (due to the high thermal capacity of water and its fluidity). Although it also provides evaporative cooling, less evaporation takes place from free water surface than by transpiration from the leaves of vegetation.

The water surface is exposed to the cold night sky, which can have an effective radiant temperature of 10° to 15°K lower than air temperature. The water surface has high emissivity and loses heat, which mixes efficiently due to downward convection. This cooled water can then be used passively, by absorbing heat conducted through the ceiling, or actively by a pumped circulation to suitable emitters.

The roof pond enjoyed some fame in the US in the 1970s, with Harold Hay's Skytherm. Moveable insulation protected the pond from solar gains in the day, when the passive conduction of gains from within were absorbed by the cool water.

Similar to the green roof, an important non-thermal consideration is the structural support. This may already be sufficient for short spans, but for large spans could involve extra investment in structural materials. Furthermore, parasitic energy used for pumping and other mechanical operations must be evaluated.

5 Windows

Windows are the most energy-transmissive elements in the envelope. Even the highest performance glazing has a U-value at least five times greater than typical insulated opaque elements. When in direct sun, windows will transmit around 400W/m², 40 times greater than a 20°C temperature difference across a wall with a U-value of 0.5W/m². As well as this, windows provide useful daylight that can reduce the dependence on artificial light, one of the main energy consumers in non-domestic buildings.

It is not surprising then, that windows and glazing have a crucial effect on building performance. Refurbishment provides the opportunity to incorporate the latest technology for glazing materials and framing components as well as adding to the 'daylight system' in the form of shading and reflecting elements.

5.1 Glazing materials

The key characteristics of glazing are the thermal transmittance (U-value) and the optical transmittance. It is the combination of these that affect the performance of glazing in admitting visible daylight (and allowing views out), and at the same time keeping the heat in, or in overheated climates, out. The balance of these two characteristics should influence the designed glazing area; increasing the area admits more light, thus reducing the need for electric light, but increases the heat transmission. It follows that the area that gives the optimum energy performance is dependent on the value of these two key parameters.

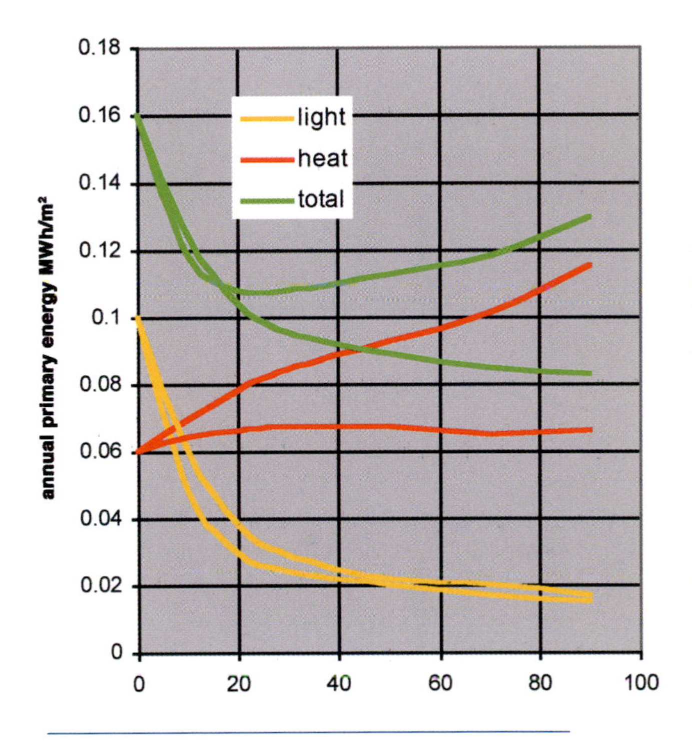

Figure 5.1 LT Curves (annual energy consumption as a function of glazing area) for different glazing types showing the effect on optimum glazing area

The physics of the energy balance of glazing is a little more complex than two characteristics alone might suggest, since heat transfer involves radiation as well as conduction and convection, and the transmission of radiation has a bearing on thermal solar gain as well as useful daylight.

Heat transmission through glazing

Figure 5.2 shows separately the processes involved in the transmission of heat and light through single glazing. Both are present simultaneously.

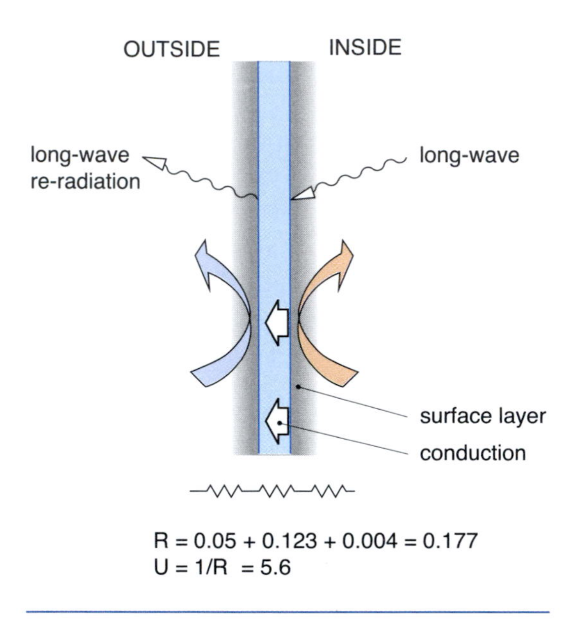

Figure 5.2 The mechanisms of energy transfer through glazing by convection and conduction

The surface of the glass presents a resistance to the convective component (in Figure 5.2) due to the stationary layer of air a few millimetres thick that acts as an insulator. This resistance is known as the surface resistance (or inversely, the surface conductance), and for still indoor air is about $0.123m^2{}^{\circ}K/W$.[1] The standard value taken for outside is lower, due to the prevalence of air movement, and is usually taken as 0.05. The resistance of the glass itself is very small, about 0.004 for 4mm glass.

The U-value is given by the inverse of the total resistance:

$$1/0.177 = 5.6W/m^2{}^{\circ}K$$

It is important to note that the majority of the thermal resistance of single glazing is due to the surface layers. If this layer is eroded by internal air movement induced by fans or convection, and/or the outer layer is reduced by wind, the overall U-value of single glazing could be increased by a factor of two. A simple explanation of double glazing is that a further two still air layers are included. However, heat transfer between the two leaves still takes place by radiation resulting in the U-value being reduced to only

Table 5.1 Typical U-values for different glazing specifications

Glazing type	U-value	Light transmittance	Total transmittance
single	5.4	0.87	0.90
double	2.8	0.76	0.82
triple	1.9	0.65	0.74
double low-e	1.8	0.64	0.58
double low-e inert gas	1.5	0.64	0.58
double tinted	2.8	0.35–0.65	0.65–0.75
double reflective	2.8	0.15–0.3	0.2–0.38
double low-e high performance	1.5	0.25–0.5	0.3–0.45

1 This includes an effective resistance to take account of radiative transfer from room surfaces to the glass.

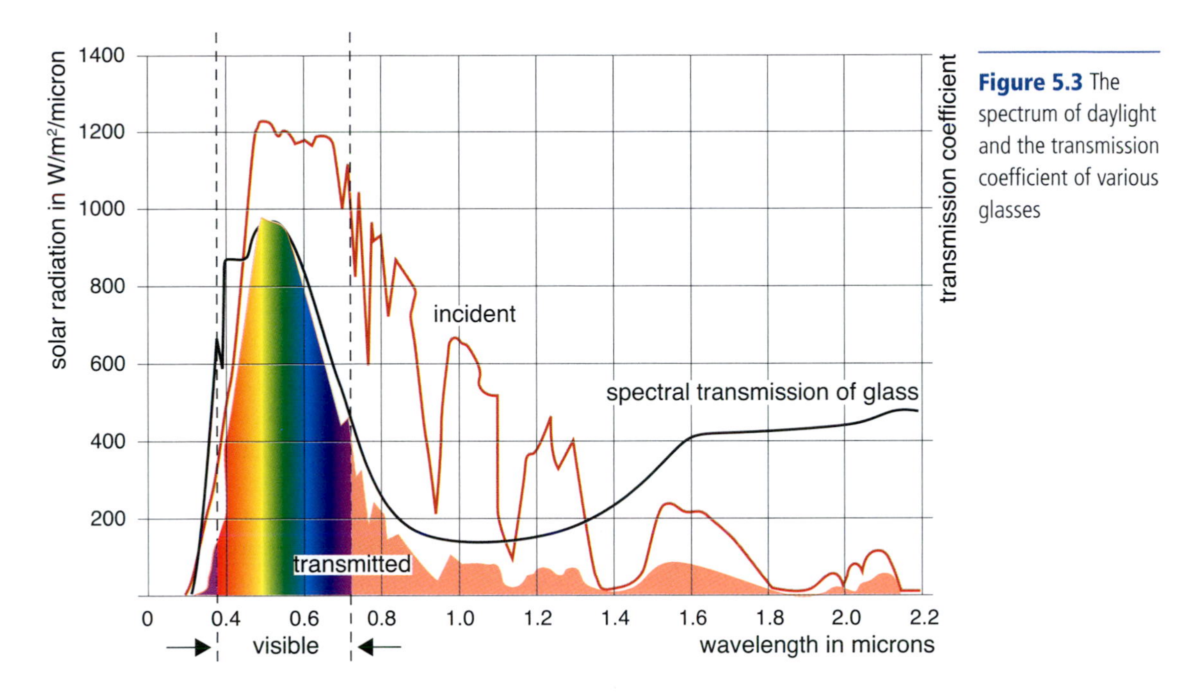

Figure 5.3 The spectrum of daylight and the transmission coefficient of various glasses

about 3W/m²°K. Note that increasing the gap between the panes beyond about 16mm does not reduce the U-value further since convection currents are set up, transferring heat from one surface layer to the other.

The next step to reduce losses is to reduce the transfer by radiation. This is achieved by coating the surface of the inner leaf that faces the cavity with a low-emissivity coating. This very thin (less than a wavelength of light) metallic layer transmits short wavelengths (i.e. visible light) but acts as a poor emitter for long-wave infrared (IR).

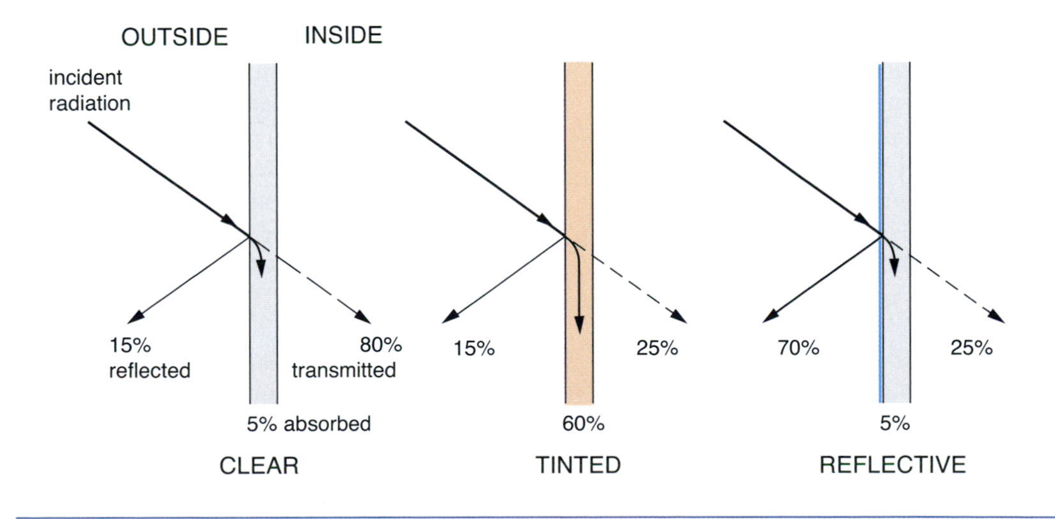

Figure 5.4 The transmission of energy by short-wave radiation through clear, tinted and reflective glass

This typically reduces the U-value to about 2.2W/m^2°K.

Finally, the convective transfer can be reduced further by reducing the conductivity of the gas in the cavity. This is achieved by replacing the air with a heavy inert gas such as argon, resulting in the U-value dropping to about 1.8W/ m^2°K.

U-values can be further reduced by triple panes and multiple coatings. Table 5.1 summarizes typical U-values for different glazing specifications.

Radiation transmission through glazing

Rather less than half the energy in daylight lies in the visible spectrum (Figure 5.3). Most of the remainder is in the infrared (IR) region and a small fraction in the ultraviolet (UV). Only the visible component is useful for daylighting. However both components are transmitted by glass and thus contribute to unwanted overheating and useful solar gain when absorbed by surfaces in the room.

Figure 5.4 shows the energy balance of the radiation incident on the glazing. The radiation falling on the glazing surface becomes three components – a reflected component I_r, a transmitted component I_t and an absorbed component I_a. For normal clear glazing, in the visible region the reflected component is about 15 per cent of the incident value but this increases strongly when the angle of incidence increases beyond about 60°. The transmitted component is about 80 per cent of the incident radiation, leaving 5 per cent absorbed by the glazing.

The proportion of transmitted, reflected and absorbed radiation varies according to wavelength, as indicated in Figure 5.3. Only a small fraction of UV light is transmitted. Roughly 80 per cent of the visible spectrum is transmitted by glass (e.g. 4mm window glass) in the visible spectrum, and about 70 per cent in the near IR region. Far IR radiation is absorbed – this leads to the 'greenhouse effect' where energy admitted as short-wave IR and visible, converted into heat, cannot then be re-radiated as long-wave IR resulting from the warming of the room's contents.

Two other types of glass are commonly encountered: tinted and reflective (Figure 5.4). Tinted or absorptive glass contains pigments to increase the absorption of visible and near IR. This reduces the transmittance, typically from 40 per cent to as low as 10 per cent. The absorbed energy heats up the glass and this heat is partly conveyed into the room and partly to the outside by radiation and convection. It is not a good solution to overheating due to excessive solar gain.

Reflective glass has a thin metallic or semiconducting coating that increases the reflected component, also reducing the transmission. However, in this case, the energy is reflected away from the glass, not absorbed by it, and thus causes no heat gains to the room.

High performance glazing

Glasses that are referred to as 'high performance' have the important property that the transmittance in the invisible spectrum (the infrared and the ultraviolet) is significantly reduced. Although the thermal gains due to the absorption of the visible light remain the same (for a given level of illuminance), gains from the invisible part of the spectrum are reduced. Thus the light can be regarded as 'cooler', or put more scientifically, to be of *higher luminous efficacy*.

This is illustrated in Figure 5.5 which shows the heat gains to a 12m^2 room for different glazing materials. In each case, the area of glazing is adjusted to give an average illuminance of 300 lux. This is also illustrated in Figure 5.6. The x-axis shows the ratio of total transmittance T_t to the visible transmittance T_v. For high performance glass, the ratio T_t/T_v is less than one. However, even if none of the invisible radiation were transmitted, this ratio could not be below about 0.5, since about half of the thermal effect of solar radiation is due to visible radiation. This is shown as the theoretical limit on the graph. Materials with ratios greater than one actually worsen the situation, since, as the graph shows,

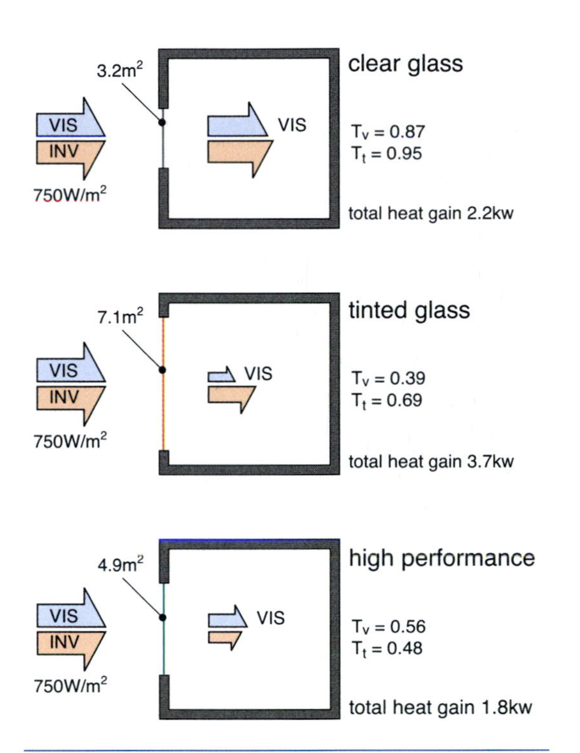

Figure 5.5 Heat gains from solar radiation for different glass types, adjusting glazing areas in order to keep the room illuminance constant

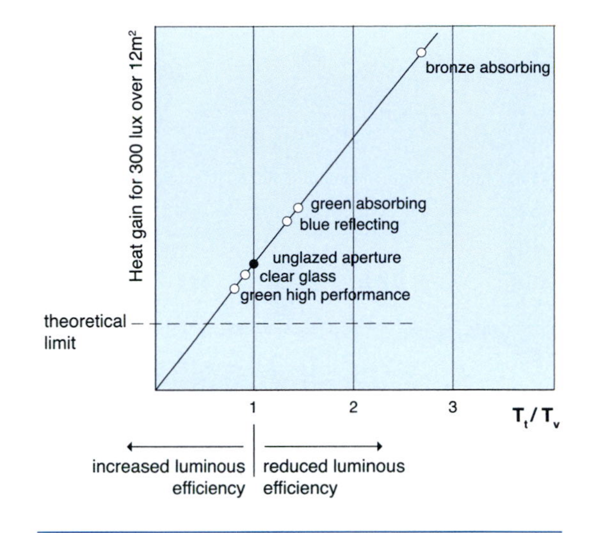

Figure 5.6 Solar heat gain plotted against the ratio of total transmittance to visible transmittance, for a given average illuminance of 300 lux

more heat is generated in the room for the same amount of light.

Buildings that have encountered overheating in the past may have had tinted glass or film retrofitted at some stage. Since the absorbed energy is partially transmitted to the room, the improvement may have been marginal. Refurbishment is an opportunity to replace with clear glazing and/or shading devices reduce the glazing area.

Table 5.1 lists generic glazing types giving their U-value, light transmittance and total transmittance. Some trade information gives the shading factor which is defined by *shading factor* $= 1 - T_T$. Note that the shading factor alone is not an indicator of performance as a window material – for example, a brick wall has a shading factor of 100 per cent!

5.2 Framing and support systems

As technology has advanced, larger and larger sheets of glass are manufactured. In new buildings this has led to supporting systems that minimize both visual and thermal impact. In older buildings the reverse is true with the framing of smaller panes of glass affecting the appearance, the light transmission and the thermal conductance of the whole window element.

An increase in the total glazing area has tended to follow growth in sheet size, which in turn has demanded novel glazing support systems, ultimately leading to the 100 per cent glass facade. Curtain walling is the principle of supporting glazing or other lightweight panels within in a lightweight frame, given lateral support from the parent structure. In the last two decades invisible fixing systems attaching to the backs of panels have developed, now making the 100 per cent glass facade possible.

Before the development of high performance glass, these systems often led to grossly overglazed buildings suffering from excessive heat

Figure 5.7 The evolution of glass pane size: top left: 17th century cottage Weald and Downland Museum, Sussex, UK; bottom left: Georgian terraced houses in London; above: 100 per cent glazing in 1990s office block, Berlin

loss in winter, with glare and unwanted solar gains in summer. These buildings are amongst those now eligible for refurbishment and solutions may well involve major intervention to the glazing.

Obstruction of light due to framing

The obstruction of the framing and glazing support system is often greater than expected, as illustrated in Figure 5.8.

The analysis here only considers the obstruction of the framing in the plane of the glazing. Framing has a third dimension and this can also have an effect on the transmission of the window at greater angles of incidence. For example, a window in a narrow street may have a sky view only at angles greater than 45–60°. Heavy mullions and transom elements when seen projected at such angles will obscure a greater area than when viewed at right angles, and if they are finished in a dark colour, will seriously reduce the daylight transmission.

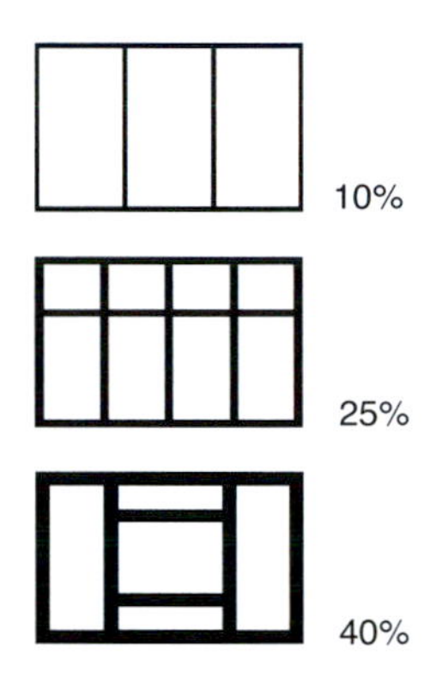

Figure 5.8 Obstruction due to framing

Refurbishment could offer an opportunity to make improvements either by reducing the size of the framing elements or, where there is a significant three-dimensional effect, increase the reflectance. Both of these measures will have a visual impact to the elevation and may therefore be precluded for historic buildings.

Thermal performance of framing

The framing of panes of glass can have a large impact on the overall U-value of the glazing plus frame. As with light transmission, it is a three-dimensional problem, not just the proportion of the opening occupied by the frame in the plane of the glass. Figure 5.9 shows a serious case where aluminium framing increases the U-value from 5.6W/°Cm^2 to an average U-value of 9.5W/°Cm^2. This is due to the 'fin effect' – the fact that the internal and external surface areas of the mullion are the limiting conductance rather than the cross section of the aluminium. This is because of the very high thermal conductance of aluminium, at 237W/m°K it is 1500 times greater than the thermal conductivity of wood. Nevertheless, the ability to extrude aluminium with highly complex sections renders it ideal for framing elements supporting glass panes and seals for opening windows. (Modern window sections have partially overcome the problem of high conductivity by introducing thermal breaks (insulated sections) in the framing components.)

Unfortunately, windows of the design and material in the example above are relatively common in buildings built in the last 50 years, and are now become candidates for refurbishment. Clearly upgrading windows of this design by replacing the glazing alone will not be sufficient in itself. In the example above, the aluminium framing increases the U-value by 70 per cent. The same construction glazed with low-e argon-filled double glazing with a 'mid-pane'[2] U-value of 1.6, would be increased to 4.8 by conduction via the aluminium mullions, an increase of 200 per cent.

There are three options available in this example. The first option is to insulate the mullions, either inside or outside. This is labour intensive and may present problems due to change of appearance. In our example, practical considerations limit the insulation thickness to about 25mm; thicker than this begins to obstruct the glazing significantly. However, this kind of glazing is often encountered in buildings that have more than sufficient glazing area, thus some obstruction, though changing the appearance, will still leave adequate glazing area, particularly with a high reflectivity finish.

The second option is to install a secondary glazed screen of high performance glazing and

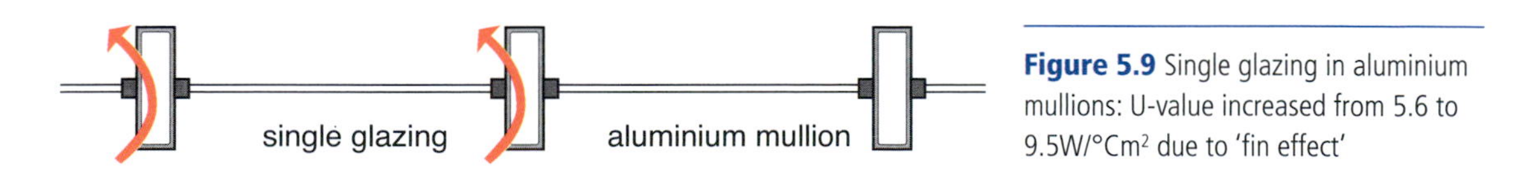

Figure 5.9 Single glazing in aluminium mullions: U-value increased from 5.6 to 9.5W/°Cm^2 due to 'fin effect'

2 U-value of pane away from the influence of framing element.

thermally insulated framing. This could be inside or outside, and carries the advantage that it could form the main airtight layer, thus addressing a second problem of air-leakage commonly found in older glazing systems. If outside, the skin can also form the weathering layer.

The extreme application of this strategy is the second skin as described in Chapter 9 and illustrated by the Albatros project in Figures 9.3 and 9.4. Clearly secondary glazing will affect the appearance, either from inside, or outside, or as in the case of the Albatros, both. And there may be high costs, particularly associated with external secondary glazing.

The third option is to replace the glazing and the framing system with high performance glazing and thermally insulated framing. This will also involve high costs in both money terms and embodied CO_2.

Framing material

Other materials commonly used for framing systems are UPVC and wood. The thermal conductivity of UPVC is about 0.2W/m°C, nearly as low as timber at 0.15W/m°C, both orders of magnitude lower than aluminium at about 237W/m°C.

UPVC also can be easily extruded into complex sections and claims are made for its low maintenance and long life. However, it is vulnerable to UV degradation, losing its glossy surface and becoming brittle over time spans of 20 years or so. PVC windows also have much greater environmental impact than timber, due to their embodied CO_2, pollutants during manufacturing and difficulty in recycling.

Timber carries none of the above disadvantages, being almost carbon neutral. Good quality timber, with good detailed design and maintenance can last for a century or more. Many historic buildings from the 19th century still have their original window frames, where building of 100 years later have already had one or even two replacements. The reasons for this long life the quality of the timber – often high-latitude grown resinous pine, spruce and fir, and the detailed design which discouraged the harbouring of water and migration of damp from masonry. Unfortunately, in an effort to 'modern-

Table 5.2 U-value of glazing and framing combinations for pane size 1.2m x 1.2m. These values are indicative only. The full procedure for calculating whole window U-values can be found in ISO 10077-1

Glazing + framing system	U-value of glazing material	U-value of glazing + framing
Double glazing Aluminium	2.8	3.1
Double glazing Aluminium with thermal break	2.8	3.0
Double glazing low-e inert gas Aluminium	1.5	2.7
Double glazing low-e inert gas Aluminium with thermal break	1.5	1.9
Double glazing low-e inert gas Timber	1.5	1.8
Double glazing low-e inert gas Timber with thermal break	1.5	1.5

ize' in style and production, many of these attributes are missing in window design from the 1950s onwards.

In spite of the long life of these early windows, because they were predominantly single glazed and were often poorly fitting, they present problems in sustainable refurbishment. The insertion of double or triple glazed panes into the slender wooden sections is virtually impossible. There are a number of products available for improving airtightness, but these also have some visual impact and it is difficult to achieve the same performance as a modern window.

The recognized historic value of the building will often set priorities here – in some cases tolerating the poor thermal performance and seeking other ways to reduce the carbon emissions of supply, by means such as combined heat and power (CHP) or renewable heat sources. In other cases, compromises may be possible such as the installation of internal secondary glazing.

Table 5.2 summarizes the effect of framing on the combined U-value of glazing and frame. Note that it is sensitive to the proportion of framing to the total area of opening in the insulated building envelope.

5.3 Modifying apertures

Many buildings from the 1950s onwards, which are now being considered for refurbishment, have large areas of glazing. In many cases this is single glazing, and would thus be a possible candidate for glazing replacement. If aesthetic considerations allow it, a reduction of the aperture area could be considered. This will have several advantages – reducing heat loss and unwanted solar gain, reducing glazing costs, reducing shading costs, and in some cases supporting interior remodelling by providing more wall space for furnishings and equipment.

Since the new opaque envelope can have U-values as low as 0.2, the impact on the average U-value can be significant. For example if the 70 per cent double glazing of a facade is reduced to 35 per cent with an opaque material with U-value of $0.2W/m^{2o}K$, the average U-value of the original aperture would be reduced from 2.16 to $1.18W/m^{2o}K$. If it was originally single glazed and was reglazed with low-e of 35 per cent glazing ratio, the average U-value would be reduced from about 3.56 to 0.83.

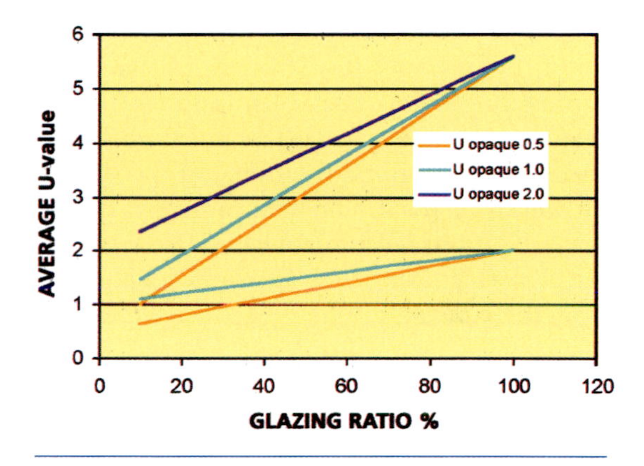

Figure 5.10 The average U-value as a function of glazing ratio

Although 35 per cent glazing can be shown to provide adequate daylighting for rooms of up to 6m deep (for unobstructed windows), care has to be taken with how the distribution of glazing is altered. General rules of thumb are for spaces with a 2.7–3m floor to ceiling height:

- Cill heights should not be raised above 1m from floor level.
- Glazing in the upper part of the wall is more effective than the lower.[3]
- Horizontal distance between glazed openings should not exceed 3m.[4]
- There may be a case for splitting the viewing and daylighting function.

3 Except when there are deep overhangs.

4 Or 2m from a cross-wall in cellular offices.

There may be cases for increasing the glazed area, where daylighting is demonstrably inadequate. However, other causes for poor daylight performance should be eliminated first – for example, low transmission of glass, obstruction due to framing or poorly designed fixed shading devices, low reflectance of interior surfaces, or internal obstructions.

There may be cases for changing the distribution of glazing by making new apertures. Single-storey buildings and top floors offer a good opportunity for this. Small areas of rooflighting over a deep plan can make a dramatic improvement in daylight distribution and, therefore, the luminous efficacy of the system. However, in warm climates, rooflights should never be unshaded horizontal glazed apertures, but always structures with apertures sloping away from the equator.

Figure 5.11 The traditional louvred shutters on the south facade retained as shading devices for the Meyer Hospital refurbishment, a REVIVAL project

5.4 Shading systems

The need for shading as part of any window system is now widely recognized. Only in the era of cheap energy and immature modernism was shading regarded as an inconvenience and an aesthetic intrusion that could be dispensed with by the substitution of air-conditioning. The enthusiasm for large areas of glazing and lightweight construction exacerbated the problem, which we have now inherited.

Vernacular architecture, with its rich tradition of louvres, shutters, jaloises, overhangs and blinds clearly demonstrates that, before this profligate period, passive low-energy design had to include these elements. With global warming and the gradual shift from the problem of heating to that of cooling, the understanding and adoption of shading has become an essential part of sustainable refurbishment.

Daylight redistribution

Shading devices have an important function other than simply reducing the quantity of radiation transmitted through the window. Due to the reflection of light from surfaces and/or the obstruction of light from specific directions, some devices can improve the distribution of daylight in a room.

Side-lit rooms from plain apertures are very unevenly lit, as shown in Figure 5.12. This means that in order that the minimum illuminance is sufficient, the part of the room near to the window has to be over-illuminated. At times of zero heating load this over-illumination serves no useful purpose, but will contribute to overheating and probably glare.

If the light distribution can be improved, the total amount of light flux entering the room (and the invisible radiation accompanying it) can be reduced. Thus the ideal shading device not only modulates the transmittance of the window, compensating for conditions of high radiation levels, but also improves the efficiency of distribution. Figure 5.12 shows how a light shelf reduces illuminance close to the window more than at the back of the room. Thus, the unwanted part of the incident radiation is reflected away, reducing solar gain. Horizontal louvres can work in a similar way.

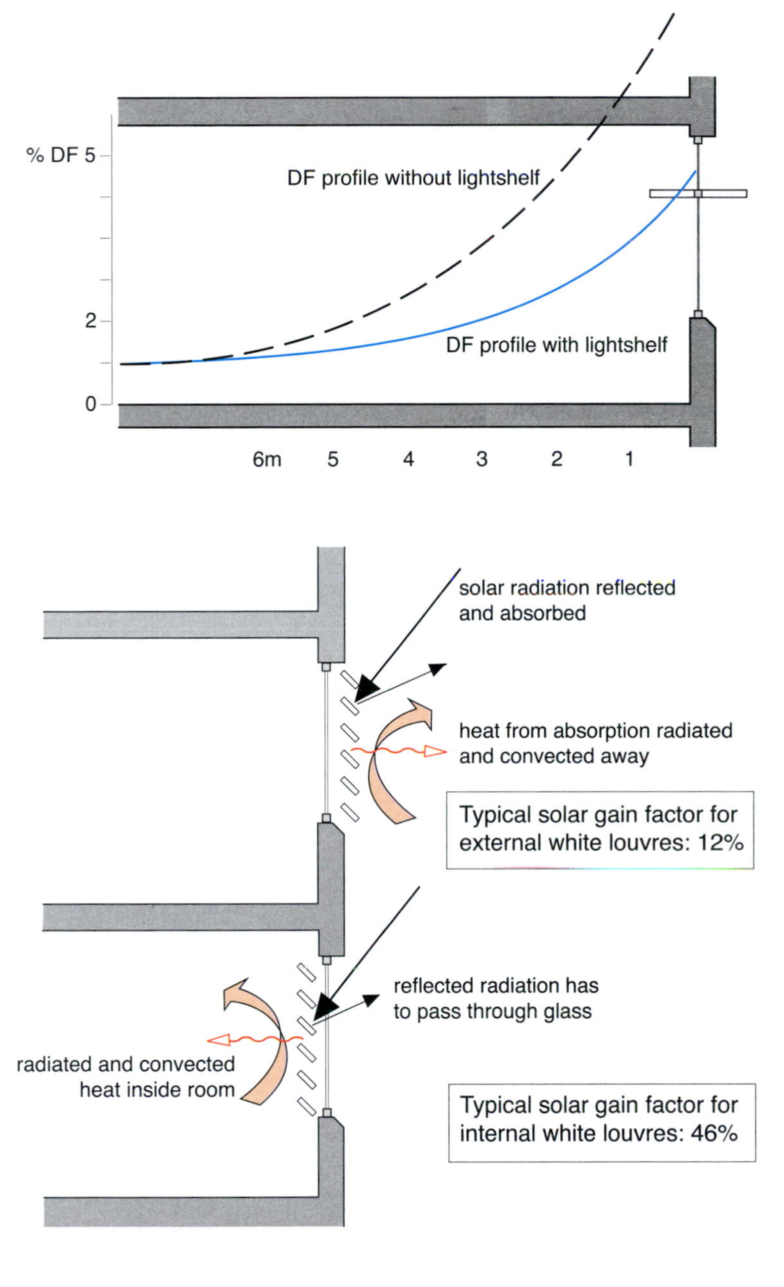

Figure 5.12 The daylight profile of a side-lit room. A light shelf can reduce the illumination close to the window without reducing it at the back of the room, thereby increasing the illumination efficiency

Figure 5.13 Comparison between external and internal shading

Shading options for refurbishment

Some of the buildings where the architectural specification has not included shading provision will have had shading retrofitted at some stage. Most commonly this will be internal curtains, blinds or louvres, but at worst, pieces of paper stuck onto the glass by desperate occupants. The installations are often poorly integrated with the window design, obstructing controls and interfering with ventilation, and are usually poorly maintained. These experiences give shading a poor reputation.

Options for shading fall into four categories[5] – *External, Internal, Interpane and Integrated*.

It is well known that the thermal performance of external shading is superior to internal shading, since re-emitted heat is lost to the outside, rather than into the room, as shown in Figure 5.13. However, internal shading is usually cheaper and easier to control.

5 These notes are intended to assist in strategic decisions about shading. Detailed design and specification should use analytical design tools to assess the performance quantitatively.

Interpane shading refers to devices held between panes, or in the cavity of a 'double skin' envelope. These systems have the advantage that the shading device is protected from weather, dust and mechanical damage. It is important that the design of the cavity permits adequate ventilation to the shading elements.

Integrated refers to devices such as lightshelves and prismatic systems which explicitly address the daylight distribution function as well as selective shading.

External shading

External shading broadly falls into the following types:

- overhangs (fixed or retractable);
- louvres (fixed, adjustable, retractable);
- fins (fixed, adjustable);
- blinds (retractable);
- perforated screens (fixed).

Shading type	Orientation	View	Nat vent (in limiting conditions)	Daylight (in limiting conditions)	Seasonal response	Modulation	Notes
overhangs							
fixed	180 +/- 30	good	good	medium	medium	none	e.g. canvas
retract	180 +/- 30	good	good	good	good	good	awnings +
							adjustable
							geometry
louvres							
fixed	180 +/- 30	med –	good	medium	medium	none	view
adjust	all	poor	good	good	good	good	influenced by
retract.	180 +/- 30	med –	good	good	good	medium	blade module
retract.+	all	poor	good	good	good	good	size & geom.
adjust		med/good					'good' applies
		med/good					to when
							retracted
fins (vertical)							
fixed	90, 270	med –	good	medium	medium	none	view
	+/- 20	poor	good	good	good	good	influenced by
adjust	90, 270	med					blade module
	+/- 45						size & geom.
blinds							
retract	all	poor/good	poor	good	good	medium	'good' applies
							to when
							retracted
perforated screens							
fixed	all	poor	med - poor	poor	poor	none	not
							recommended

> ***fixed*** – fixed
>
> ***adjustable*** – remains in position but radiation transmission can be modulated (e.g. altering angle of louvre)
>
> ***retractable*** – can be removed completely from the aperture

A summary of their properties is given in Table 5.3. When using the table please note the following:

- Orientation *all* implies that performance is not orientation-sensitive, although in general, there would be no demand for shading on facades orientated towards the poles +/–45°.
- Adjustable louvres and fins will have poor *view* performance and poor *natural ventilation* performance, when completely closed.
- *Natural ventilation limiting conditions* refers to situations requiring the shading to be deployed, that is when there is an overheating risk. *Good* implies that natural ventilation is not impeded.
- *Daylight limiting conditions* refers to minimal daylight availability when there would be no requirement for shading. *Good* implies that the shading devices cause no reduction in daylight transmission in these conditions.
- *Distribution* refers to the ability of the shading device to improve daylight distribution, thereby lowering the total daylight required.
- *Seasonal response* refers to the ability of the shading device to respond to different sun angles at different seasons.

Internal shading

This is often the choice for retrofit. Options are limited to louvres (venetian blinds) and roller blinds of translucent or opaque material.

Horizontal louvres

These offer some possibility for improved light distribution by using a combination of high and low reflectance finishes. If the upper surface is light, and the lower surface somewhat darker, light will tend to be directed upwards towards the ceiling, thereby improving the illumination at the back of the room. Some proprietary systems use specially shaped louvres with specular (mirror) reflecting surfaces, in order to further improve the performance.

Most internal louvre systems are retractable, allowing maximum daylight transmission at times of limited daylight availability. This also permits cleaning. Whilst open louvres can allow a good flow of air, they should be anchored top and bottom to prevent movement and noise. However, access to window opening controls must be provided. These conflicts often result in low cost simple installations being of poor functional quality, as shown in Figure 5.14.

Figure 5.14 Office in Athens where internal louvres interfere with the opening windows and ventilation. A REVIVAL project before refurbishment

Roller blinds

These are another common retrofit solution due to their low cost and ease of installation. Consideration must be given to the optical properties of the fabric as indicated in Figure 5.15.

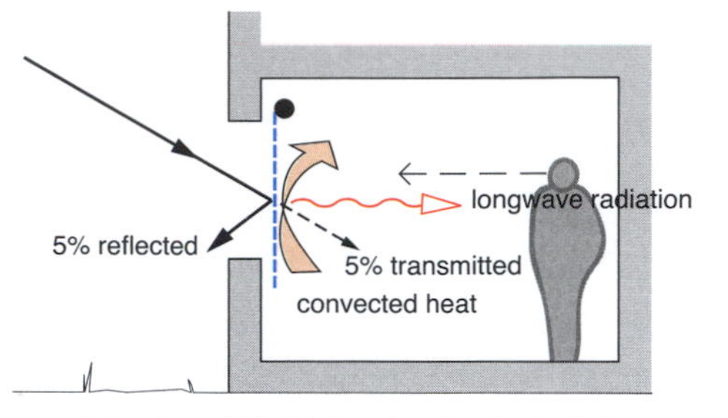

dark coloured blind: internal surface looks dark

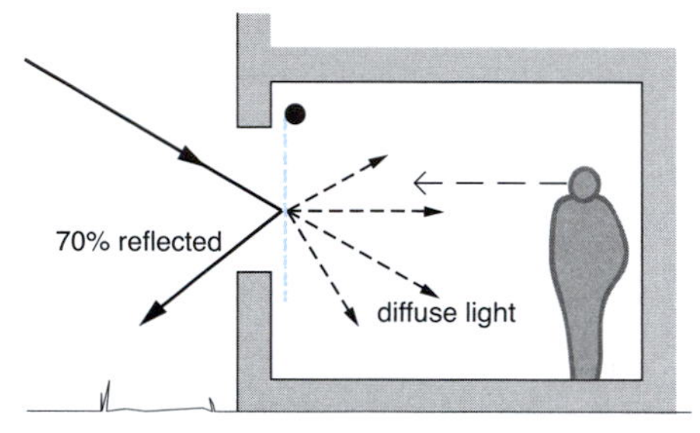

light coloured blind: internal surface looks very bright

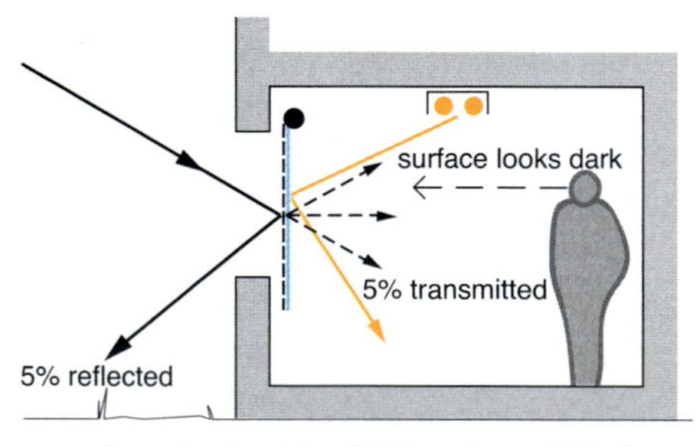

externally aluminized light coloured blind:
internal surface looks dark for external light but
reflective to internal artificial light

Figure 5.15 Optical properties of blind materials

Roller blind fabrics are usually optically diffusing, which means that light is re-emitted in all directions. This means that blinds of moderately high light transmission could become glare sources themselves when illuminated by direct sun.

Blinds with an aluminized surface on the outside have a good performance, reflecting both visible and near IR radiation. If the aluminium coating is applied to 100 per cent of the surface, no light will be transmitted, allowing total blackout to be achieved. However, the internal surface facing the room can still be white or light coloured, providing improved lighting performance at night under artificial lighting, compared with a black or dark coloured blind. Blinds with partial aluminizing on the outside surface will show improved shading performance and also present a moderately luminous surface inside. Both types have reduced heat loss at night due to the low emissivity of the metallic surface.

When roller blinds are used in conjunction with overhangs, consideration should be given to positioning them so that they deploy upwards – that is, with the roller at cill level. This means that sunlight striking the lower part of the window beneath the overhang will be intercepted, whilst allowing light and view through the upper part. Alternatively, the lower part of the window can be shaded separately.

5.5 High performance daylighting[6]

High performance daylighting is an integrated approach that takes all the functions of the window system and room into account. The principle is to maximize the benefit of daylight and to minimize the disbenefit. This involves three actions:

1. Increasing the luminous efficacy of the light (reducing its heating effect) by the specification of the glazing.
2. Improving the spatial distribution of light in the room to reduce over-illumination.

6 This topic is dealt with in more detail in Revival Technical Monograph 4: High Performance Daylighting, available on the website www.revival-eu.net.

3 Responding to the varying availability of light from the sky diurnally and seasonally to prevent over- or under-illumination.

In most cases, large improvements can be made with glazing and shading elements, and control systems, without necessitating major structural alterations. This makes high performance daylighting an appropriate objective for refurbishment projects.

6 Atria and Double Skins

Although originally referring to the open-to-the-sky central court in classical Roman houses, the atrium was reinterpreted in the 1970s and 1980s to mean a large prestigious glazed space, usually part of a non-domestic building. Its association with energy conservation possibly stemmed from work carried out in Cambridge in the 1970s, where the energy implications of covering over courtyards was analysed, with the conclusion that there was indeed a *potential* for energy saving, provided the space was not heated. There then followed an epidemic of atria all claiming to be energy saving, although in almost all cases, the lighting, heating and often cooling of these spaces, together with the reduced availability of daylight and natural ventilation to adjacent fully occupied spaces, actually meant that these features were costing extra energy, rather than saving it. In spite of this, due no doubt to the architectural potential of these spaces, the atrium continues to be popular.

The double skin has a slightly different origin. Although historic precedents exist from the 1960s, in particular the innovative Wallesey School (Figure 6.1), the term has become widely used only in the last two decades. The continued love affair between architects and glass was being strained by calls for better energy performance than could be delivered by a conventionally skinned curtain-walled building, even with the improved thermal performance of low-e double glazing units.

Double skins, distinguished from double-glazing by the use of a gap between the layers of anything from 0.3 to 1.5 metres, not only offered improved thermal resistance, but also provided a location for shading devices, and possibilities relating to ventilation exchange between the outside, the void and the building. The *intelligent skin* – an envelope that responded to changing environmental conditions – had long been a dream of architects; now, with relatively conventional materials and a bit of help from IT-driven 'intelligent' controls, it was achievable. However, rather like atria, few double skin buildings have

Figure 6.1 Wallesey School by Morgan in 1967, set out to be self sufficient in energy, combining a double skin south facade with highly insulated thermal mass

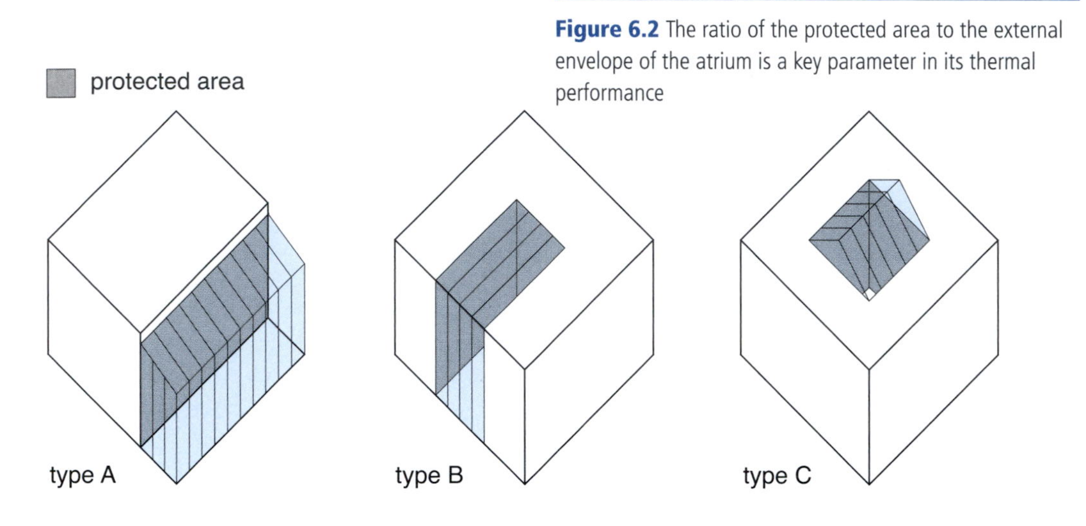

Figure 6.2 The ratio of the protected area to the external envelope of the atrium is a key parameter in its thermal performance

been shown by monitored results to have dramatic improvements in energy performance. Furthermore, it is difficult for theoretical analyses to show an advantage over a conventional partially glazed, well-insulated building, unless optimistic assumptions are made.

However, if the principle of the atrium remaining an unserviced space is adopted, then atria do have a potential for reducing energy consumption, particularly in the refurbishment of existing buildings. Similarly, there is a role for the application of a second skin to an existing

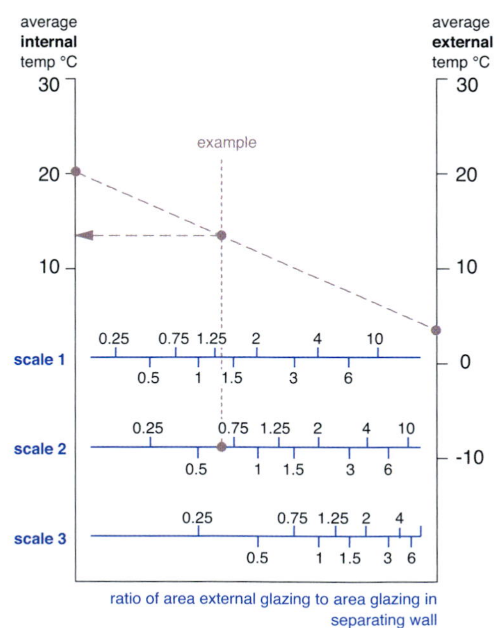

- Draw a straight line between appropriate points on the internal and external temperature scales.
- Choose scale according to glazing type.
- Draw vertical line from ratio of glazing areas.
- From intersection move horizontally to either temperature scale for atrium temperature.

Figure 6.3 Nomogram for predicting the average temperature

building. In both cases this stems from the poor environmental performance of the existing envelope.

6.1 Atria and energy: Principles

Thermal performance

The concept of the atrium stemmed from the idea of enclosing an otherwise open space, thereby 'improving' its environment. This would not only reduce the heat losses through that part of the envelope protected, but would also increase the utility of the outside space.

The impact of the atrium is very dependent upon the geometry of the situation. Consider the 2 cases in Figure 6.2. In case A the area of external envelope of the atrium is at least twice that of the area of the parent building that is protected – referred to as the *separating wall*. In case B, the reverse is true. We say that the atrium in case B has a higher *protectivity* than in case A. Rather than areas, the key parameter is actually the ratio of the overall conductance of the separating wall to the external atrium envelope. Figure 6.3 shows a nomogram for calculating the mean atrium temperature as a function of this parameter.

The temperature will also be strongly influenced by the availability of solar gains. This also depends on the solar geometry of the atrium envelope, namely the amount and inclination of south-facing glazing, compared with 'non-solar' glazing. This is referred to as the *solarity* of the atrium. The two parameters, protectivity and solarity, can combine to give widely differing performance.

Winter performance

Figure 6.4 shows the monthly average winter temperatures for an atrium in UK climate. Since the temperature in the atrium is variable, it is appropriate to refer to it as a climate. Hence its impact can be described as a *climatic shift*. Note the effect of the two values of conductance of

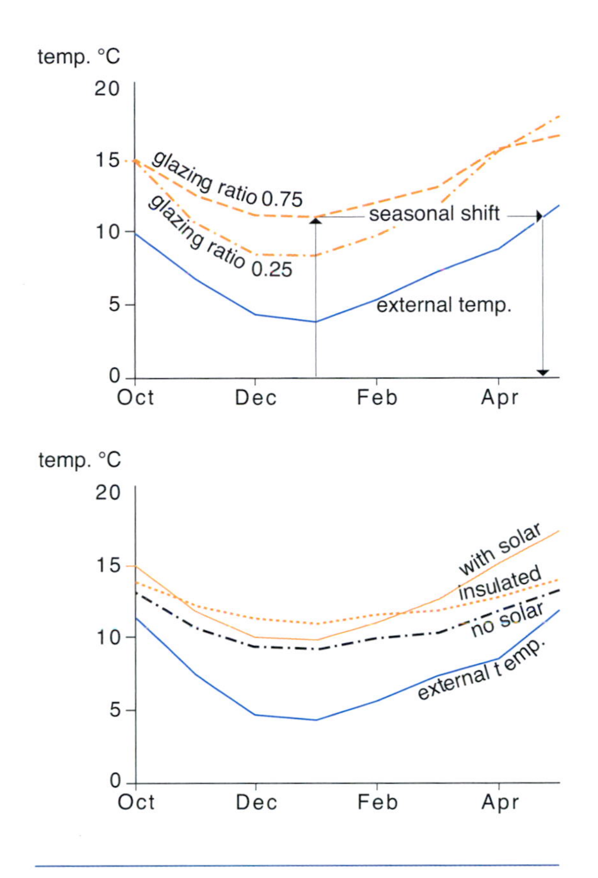

Figure 6.4 Predicted monthly temperatures for a typical unheated atrium, for different glazing ratios (R) of the separating wall, for sunny and non-sunny days, and replacing the atrium glazing with insulated opaque envelope

the separating wall – here denoted by 75 per cent and 25 per cent glazing ratio, and of the effect of solar gains.

Clearly temperatures in the daytime will be higher than the monthly average since the occupied building will be heated, and the solar gains will all be present during the day. This is indicated by the simulated results shown in Figure 6.5, and again in Figure 6.6.

The increased temperature of the atrium will obviously reduce heat loss via the separating wall, during the heating season, and hence heating load. If it is possible to draw ventilation air from

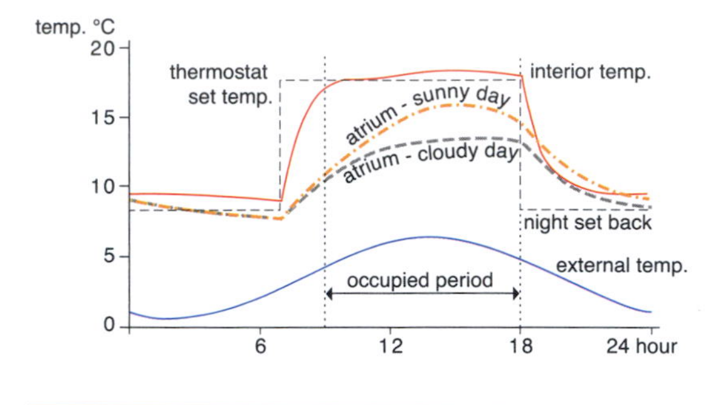

Figure 6.5 Predicted hourly temperatures of a typical atrium

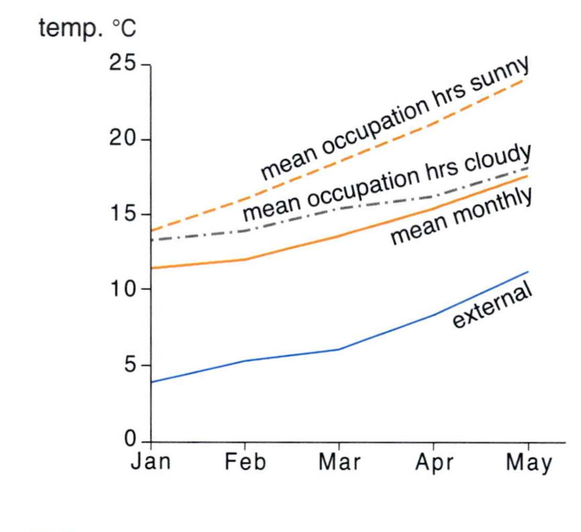

Figure 6.6 Daytime temperature increment over monthly average

the atrium instead of the outside, the ventilation heat load will also be reduced. This can lead to dramatic reductions in heat load, as illustrated in Figure 6.7.

In most real situations it will not be possible to draw all the ventilation air from the atrium, at least without mechanical ventilation. Openable windows in the separating wall will probably result in a mix of the ventilation modes, in varying proportions. Something of particular relevance

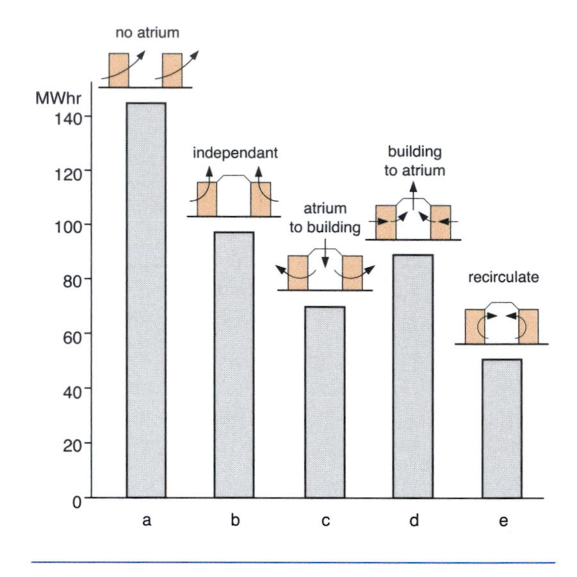

Figure 6.7 Annual heating energy consumption for a building with an atrium showing the impact of ventilation pre-heating

to retrofit is where the existing wall becomes the separating wall, and this part of the envelope can be left leaky to encourage air exchange. Recirculation alone will not provide fresh air into the system, so the atrium must be provided with fresh air, and air should be monitored in the occupied building to ensure sufficient air quality.

Please note, these temperatures and energy consumption figures are for a south-facing atrium of type B, and are indicative only. Due to the variability of the performance of atria, it is recommended that detailed thermal analysis is carried out to verify the beneficial effects of a proposed intervention, if present.

Winter ventilation to the atrium allows the supply of fresh air to the parent building if ventilation pre-heat mode is adopted. Unless the atrium is densely occupied, it will not in itself demand high rates of ventilation. Figure 6.8 shows the use of high level openings which causes good mixing, and avoids stratification.

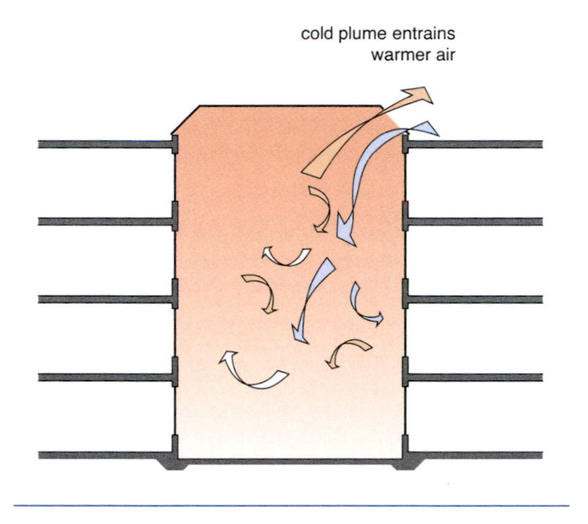

Figure 6.8 High level ventilation for winter provides mixing as the cold plume descends to floor level

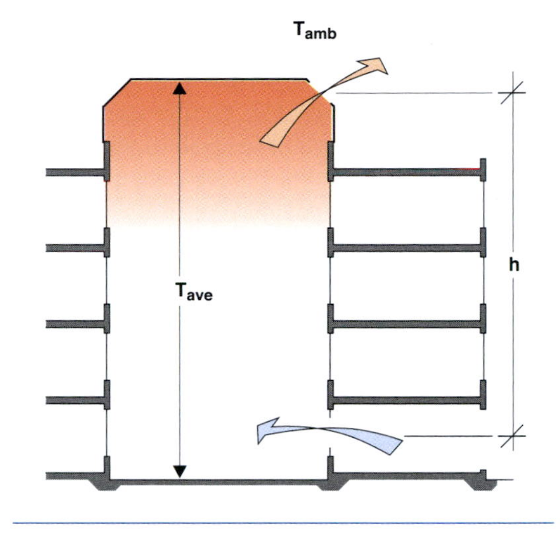

Figure 6.9 Ventilation top and bottom encourages maximum ventilation for summer. Stratification ensures that the atrium is coolest at floor level

Summer performance

The benefit of increased temperature in winter becomes disbenefit in summer. Not only may the increased temperature render the atrium space itself too hot for comfort but the parent building can suffer from both conductive heat gains and gains from ventilation. Stratification of temperature may mean that even if the atrium is comfortable at ground level, high temperatures at the top of the atrium may be transferred to the interior surrounding spaces (Figure 6.9). So the priority shifts to keeping gains to a minimum and ventilation at a maximum.

This can be achieved by adequate ventilation, preferably with inlets top and bottom to encourage ventilation by stack effect during periods of low windspeed. As a rule of thumb the openable area should be no less than 10 per cent of the 'solar' glazing area. The glass exposed to solar radiation should be shaded with moveable shading, especially horizontal or shallow inclined glazing which receives most radiation in the summer. A schematic diagram in Figure 6.10 illustrates desirable characteristics of an internal shading system. The general characteristics of shading types are dealt with in Chapter 5.

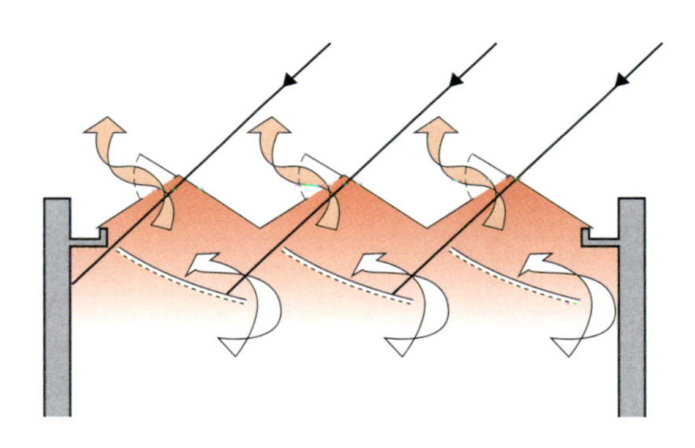

Figure 6.10 Shading and ventilation to prevent summer overheating

6.2 Effect on daylighting

In almost all cases the presence of an atrium will reduce the availability of daylight to the surrounding rooms. This is due to the absorption due to the glazing material and obstruction of the supporting structure. This can easily be a reduction of as much as 50 per cent.

It may be that some compensation can be made by adopting very light coloured (reflective)

finishes for the atrium floor and opaque parts of the separating wall. Or it may be possible to increase the glazing area in the most obstructed part of the separating wall. In any event, fixed shading and reduced transmission glazing to the atrium envelope should be avoided, unless the existing building is already grossly over-glazed. This issue should be investigated by calculating a corrected daylight factor for the occupied rooms, and using this to predict daylight availability from sky luminance data.

6.3 Planting and vegetation

Plants can use energy! Plants occupy the spaces all day every day, so any conditioning (e.g. heat or light) has to be maintained for long periods. If energy saving is a priority, it is essential to chose plants that can tolerate a cool season, and probably relatively low light levels, and require no or minimal conditioning. Even the mode of watering can have an indirect effect; watering by spraying leads to high humidity, and condensation on the glass. The response to this in cool climates is often warm-air curtains and hence a large consumption of energy. Watering should be piped to the roots and electronically controlled to avoid wasteful evaporation. In cool climates, deciduous species should be considered as these will cause less obstruction to daylight in the winter season. Avoid green tinted glass for the atrium glazing; plants do not like green light – that is why they reflect it.

Figure 6.11 This exotically planted atrium required permanent artificial lighting due to the choice of green glass for the envelope

6.4 Double skins and energy

The double skin can be looked upon as a special case of the atrium, where all (or part of) the envelope has become the separating wall, and the atrium space has become an unoccupied cavity. A double envelope on all surfaces, including the roof, is unusual; typically one of more major glazed facades will be treated.

The thermal benefits are similar – reduced conductive and infiltration losses, with the potential for ventilation pre-heating. But the disbenefits are there too, with overheating of the cavity being the largest threat. Because, typically, double envelope buildings have highly glazed envelopes, the building is more closely coupled with the conditions in the cavity, making control of these conditions, by shading and ventilation, of vital importance.

For this reason, heavy reliance is often made on 'intelligent controls'. However, both the control software and the hardware (e.g. the mechanical actuators) are notorious for error and mechanical failures. Wear and tear and erratic maintenance schedules mean that these buildings may not age well, and cannot be seen as robust design.

6.5 Other environmental factors

Open sky has an acoustic absorption coefficient of 100 per cent at all frequencies. As soon as a space is enclosed, this absorption is lost and the reverberation time is increased. This can become troublesome both for speech intelligibility within the atrium, and the propagation of noise from sources in the atrium to the inside of the building, or even from inside the building to other parts, via the atrium.

Treating opaque surfaces with acoustic absorbers can help reduce the problem. Walls, shading devices, opaque roof sections can be treated; materials such as timber strips covering mineral wool can provide good mid-frequency absorption and still maintain an appearance of an external finish. There are also micro-porous mineral materials that have similar properties. The acoustic properties of vegetation are often exaggerated, but a heavy planting scheme will definitely make some improvement. Reverberation time, and its interference on speech, are easily calculated from published materials data, and it is recommended that this analysis be done.

Flanking sound transmission, from source to receiver both located inside the building, via the cavity, is more typical of double skin buildings. This will definitely occur if two separate spaces open onto a common cavity. However the effect can be mitigated, as before, by including sound absorption within the cavity. This may be in the form of surface treatments, or provided in part by shading devices, usually located in the cavity. In these cases, the reflective path of the sound should be considered when deciding which surfaces to treat, in order to optimize its effect.

Other considerations – for example, fire-compartmentation and ventilation control – may result in the specification of separated cavities, that is each window or group of windows from one room, opening onto a single cavity. This will largely solve the acoustic problem provided the separating elements have reasonable sound insulating performance. For all of these reasons, the separating elements should be well sealed.

6.6 Atria and double skins as part of sustainable refurbishment

Most of the above applies to both new and retrofit atria or second skins. Here we consider issues particularly related to a retrofit atrium or second skin as part of a sustainable refurbishment project. Table 6.1 shows the advantages and disadvantages for atria and double skins. It should be used when considering adopting either strategy in relation to particular characteristics of the original building.

Summary

Many of the advantages and disadvantages are linked – for example, a second skin offers some acoustic protection, but it is lessened by the need to ventilate the cavity in summer; an atrium offers some reduction in heat load, but it may be smaller than the extra lighting energy to compensate for the reduction in daylight. This makes generalizations difficult, since it is only after detailed analysis, which may involve both quantifiable and non-quantifiable performance, that the net benefit or otherwise of these interventions can be evaluated. However there are some strategic rules that are reasonably robust:

- An atrium is likely to be most effective in cool climates, and when the geometry of the existing building gives the opportunity for an atrium of good protectivity and good solarity.
- In warm climates solarity may lead to overheating; shading and large openable areas for ventilation are essential.
- It will be most cost effective when the parent building provides the primary support structure for the atrium envelope.
- A double skin will be most cost effective when the costly upgrading of the original envelope can be avoided. It has the largest energy impact when the original envelope is poorly insulated and leaky. It follows that it has a larger effect on facades that are already highly glazed. For a partially glazed facade it is probably more effective to upgrade the opaque envelope and the glazed areas separately.
- For facades exposed to direct sun it is essential that the double skin incorporates shading and ventilation, and a control system that allows, but does not rely on, human intervention.

Table 6.1 The advantages and disadvantages of atria and double skins

	Advantages	Disadvantages
Atrium	Significant heating energy reduction if geometry provides good *protectivity* and good *solarity*.	Does not give significant energy savings if *protectivity* and *solarity* are *poor*.
	May offer further energy saving by ventilation pre-heating.	Reduces availability of daylight to rooms and may thereby lead to increased energy consumption.
	Can provide useful semi-climatized space at low capital and running cost.	Reduces the potential for natural ventilation by openable windows due to • propagating noise • reducing windspeeds • delivering overheated air in summer • need for fire-compartmentation.
	Provides strong architectural element making the benefits to the user of the refurbishment very evident.	The space may overheat in summer reducing its utility, and may also cause overheating in adjacent spaces.
	Will probably mean that the upgrading of the original envelope that becomes the separating wall, can be to a far lower standard than if it continued to be external envelope.	If heated, cooled and lit, the extra energy demand will almost definitely outweigh the thermal energy saved.
	May assist in stack-driven extract ventilation for surrounding spaces.	
Double skin	Significant heating energy reduction when applied to very poorly insulated envelope.	Small heating energy reduction when applied to moderate or well-insulated envelope.
	Significant reduction in infiltration when applied to leaky envelope.	May lead to overheating of parent building on E, S and W facades.
	Provides location for active shading system to original glazed facade.	Large surface area of glazing, shading, and support structure incurs high costs.
	Can provide some noise attenuation from external noise, but much reduced if cavity is open for ventilation.	Reduces the potential for natural ventilation by openable windows due to • propagating noise • reducing windspeeds • delivering overheated air in summer • need for fire-compartmentation.
	Provides architectural opportunity to change and upgrade appearance of the building.	Leads to separation from outside and disruption of views for occupants.

Final note of caution: Atria and double skins do not automatically save energy; in many cases they will lead to an increase in energy consumption and/or comfort problems. Double skins consume a large amount of embodied energy for no increase in floor space.

7 Mechanical Services and Controls

Mechanical services and controls have a large impact on the energy performance of a building, typically as much as the fabric. Clearly, the potential for improvement will depend on the current state of the mechanical equipment, which typically is updated more frequently than the building fabric. However, even recent services upgrades may not have attended in particular to improving energy performance. For example, it is not uncommon to see new low energy light fittings providing higher than necessary illuminance levels, and without controls, where no saving or even increased energy consumption is the result.

It is difficult to generalize but improvements in plant will tend to be more cost effective than improvements to the fabric. Updating, or installing controls for the first time, is even more likely to be the most cost effective strategy. With the steady development of IT and its cost reduction, large savings can be made for modest or even no extra cost, resulting in a rapid payback. When considering embodied CO_2, the payback is even more rapid since the material content of control equipment is very small.

7.1 Boilers

Boiler efficiency is the ratio of useful heat output to the calorific value of the fuel input. Boilers are required work at a range of outputs in response to the varying seasonal heat load. A characteristic of older boilers with heavyweight heat exchangers was reduced efficiency at part load, often as low as 45 per cent at one-third full load. Since full load operation occurred relatively rarely, the seasonal efficiency was considerably reduced from the instantaneous efficiency, typically as low as 55 per cent. Modern boilers have addressed this problem with improved heat exchangers and control and can return seasonal efficiencies as high as 85 per cent. Other improvements include improved combustion, and in the case of condensing boilers, the latent heat that would otherwise be lost with the water vapour, a combustion product, is reclaimed from the flue gas by condensation.

Condensing boilers can give the highest efficiency if the return water temperature can be kept below 55°C for prolonged periods. This is possible if emitters, such as fan coil units or underfloor systems, are sized appropriately, (i.e. larger than for high water temperatures) and if a controller is used that reduces flow temperature when demand is low, but allows higher temperature for warm-up and very cold periods.

It is common to use multiple boilers so that individual units can be fired up as necessary, maximizing the number of boilers working at full load and thus their highest efficiency. This arrangement also gives the opportunity to use a combined heat and power (CHP) system, or condensing boiler as the lead boiler, with less efficient (but possibly cheaper) boilers brought into use at times of peak demand only.

Separate boilers for space and (centralized) sanitary hot water can be used and this may be more efficient where annual demand profiles are very different, or where small quantities of hot water are required for summer periods.

7.2 Heat distribution

Water

Distribution systems can be wasteful in three ways:

1. Zoning (of pipework) does not allow the spatial matching of demand with delivery.
2. Pipe runs are poorly or non-insulated even when losses are not within the heated envelope.
3. Pipe diameters are too small, incurring high pump energy use.

Modest interventions to the pipework could improve zoning, and together with valve actuators could be part of a general system upgrade. It will be particularly relevant when use patterns in the building are changing, and maintaining the original zoning would lead to wasteful heating. The appropriate zone/time controls are obviously an essential part of the improvement.

The retrofitting of pipe insulation will be possible in most cases. Special attention should be paid to outdoor distribution mains, where heat losses do not contribute to useful heating, and the effects of weather and ingress of water may have severely degraded the original insulation. It is likely that older pipe insulation will contain asbestos, and this alone will probably justify its replacement on health grounds.

Small pipe diameters will lead to higher energy consumption by the circulation pumps. It may be possible to reduce flow speed due to rezoning. It may also be possible to reduce flow rates in response to reduced heat demand due to insulation improvements. Where this is not possible, replacing existing pipework with that of large diameter could be justified; for a given flow rate, the energy consumed is inversely proportional to the fourth power of the pipe diameter.

However, it is not only the pipework causing flow resistance, but often components such as valves, junctions and bends, and the resistance of these should be minimized. Furthermore, if flow rate is being modulated by constrictor valves, significant savings can be made by the use of variable flow-rate pumps.

Air

Similar considerations apply to air ducting where it is part of a warm-air heating system or full air-conditioning. Losses or gains from between the duct and ambient air can be significant. The condition of the duct insulation must be assessed and the specification upgraded to bring it to at least current standards.

CO_2 emission from electrical fan energy can be as great as that from the heat production in the boiler (Figure 7.1). This is influenced by the resistance of the ductwork and the flow rate. Again it is less likely that the ductwork will be increased in size, but it is possible that air velocities can be reduced in response to lowered demand, and variable volume fans installed into existing ductwork will allow demand for heat, coolth or fresh air to be more closely matched with occupancy. This could make significant savings in fan energy. Variable speed drives for

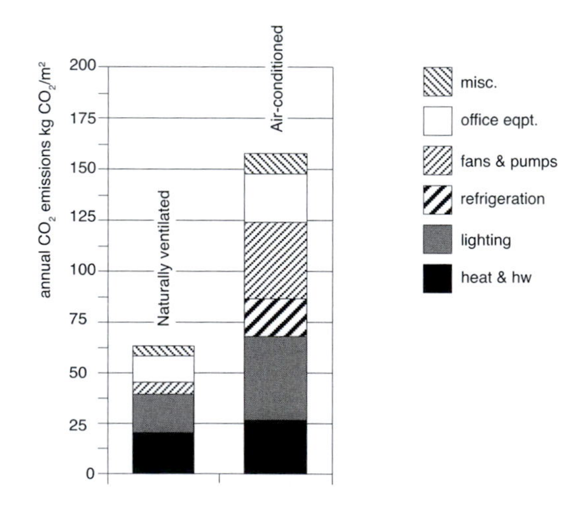

Figure 7.1 CO_2 emissions by energy end use by 14 office buildings in the UK

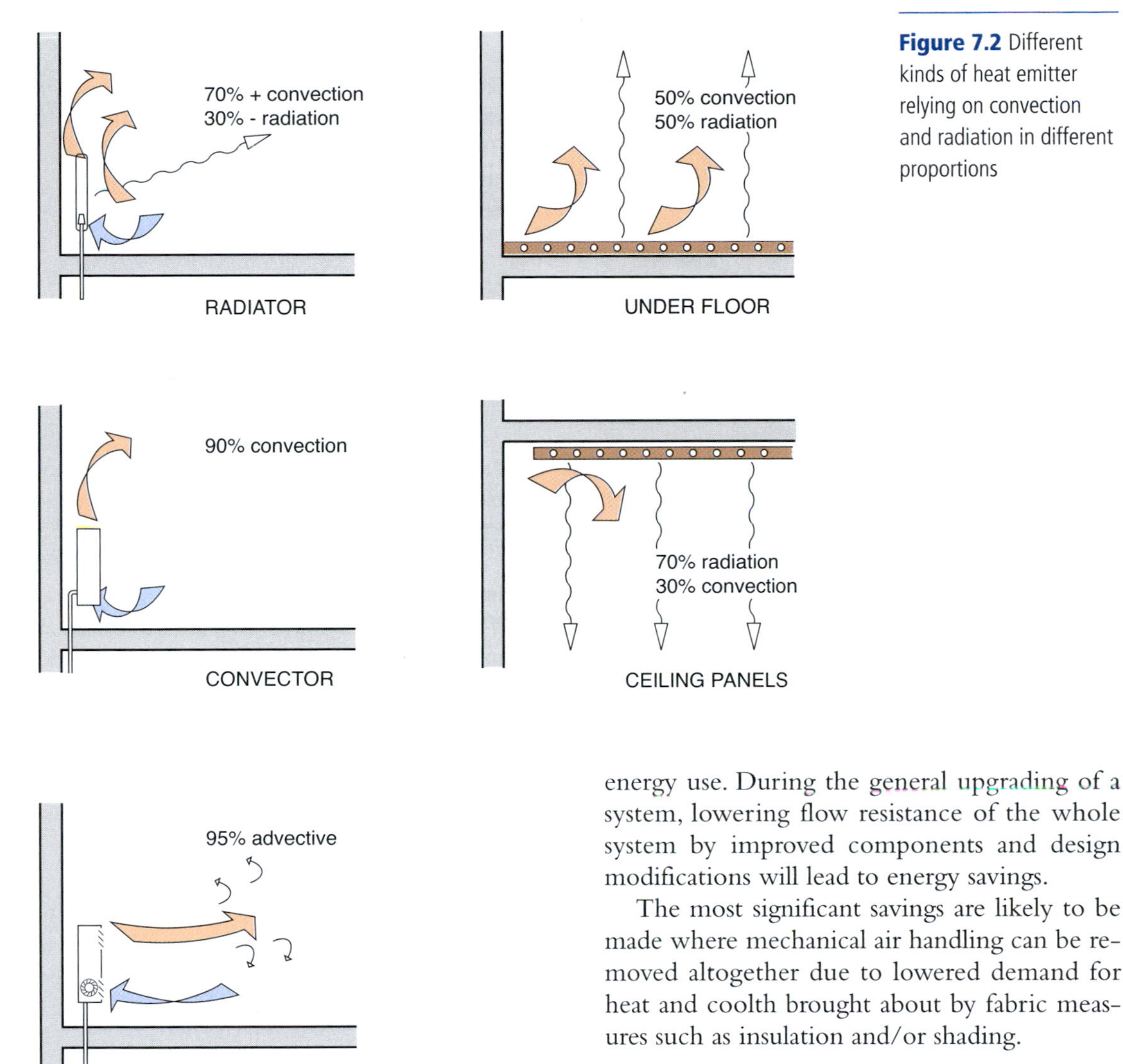

Figure 7.2 Different kinds of heat emitter relying on convection and radiation in different proportions

ventilation fans controlled by CO_2 sensors are very effective at reducing fan energy use. See section 7.4.

Closely related to duct size and the resulting flow resistance, is the resistance of air-conditioning components such as noise attenuators, humidifiers and filters. Filters are frequently neglected and pressure build-up due to clogging will activate a by-pass, thereby delivering unfiltered air, or increase the flow resistance and increase energy use. During the general upgrading of a system, lowering flow resistance of the whole system by improved components and design modifications will lead to energy savings.

The most significant savings are likely to be made where mechanical air handling can be removed altogether due to lowered demand for heat and coolth brought about by fabric measures such as insulation and/or shading.

7.3 Heat emitters

Heat is emitted into the room by one of six methods (Figure 7.2):

1 radiators;
2 convectors;
3 fan coils;
4 underfloor heating;
5 high level radiant panels;
6 via the ventilation air.

It is well known that the conventional radiator is predominantly a convector with more than half of the energy being emitted by natural convection. Multiple panel radiators (two or three layers) have a higher proportion of convective output, and the convector simply takes this further by increasing the heated surface area, usually enclosed by a casing. Fan coils takes the principle one more step by forced air flow ensuring that more than 95 per cent of the heat output is to the air. All of these emitters have to be located against walls, usually perimeter walls where the greatest heat loss is located.

Underfloor heating is achieved by heating pipes in the screed or under the deck as described in section 2.2. It introduces heat over the whole area of the floor (or the zone in which the pipes are located) with about 50 per cent as radiant energy and 50 per cent convective.

Radiant panels in the ceiling, if horizontal, have the highest proportion of radiant energy for any low temperature emitter. This is because upward-flowing convection currents cannot be set up, resulting in a higher surface temperature and hence higher downward radiant loss.

The location of the emitter and its radiant/convective split influences the temperature distribution, in particular in the warm-up period. Consider the cases in Figure 7.3. The vertical warm-air plume from the radiator or convector has to 'fill up' the room from the top down before the air temperature increases at floor level. This will lead to high temperatures under the ceiling, possibly exacerbating roof losses or ventilation losses from high level openings.

However, if the ceiling is very lightweight, this will warm quickly and begin radiating downwards. Air leaves the fan coil unit with much more momentum than the simple convector and can be directed horizontally ensuring rapid mixing.

The underfloor system increases the radiant temperature and the air temperature over the whole floor surface. This can be rapid in the case of the lightweight deck, or rather slowly when the heating pipes are located in a screed, due to the thermal inertia of the heavy material. Horizontal ceiling radiant panels radiate downwards increasing the radiant temperature and warming the floor.

These dynamic and spatial differences can have considerable influence on energy consumption, and should be considered when replacing heat emitters. For example a large double-height space of heavyweight construction would respond very badly to perimeter radiators or convectors, whereas smaller spaces with a lightweight well-insulated ceiling, will be effectively heated by such a system. Fan units carry the advantage of a high output for small size, and the ability to cause good mixing and fast response; however, the fans use electrical power, make noise and require maintenance.

Thermal comfort can be achieved at relatively low air temperatures provided the mean radiant temperature[1] is higher. Overhead radiant panels

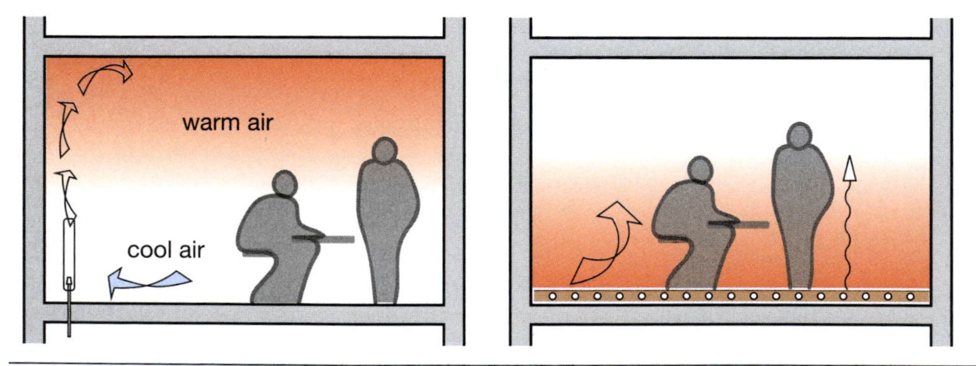

Figure 7.3 Transient effects during warm-up period for different emitters

1 The average effective temperature of all surfaces surrounding the occupant.

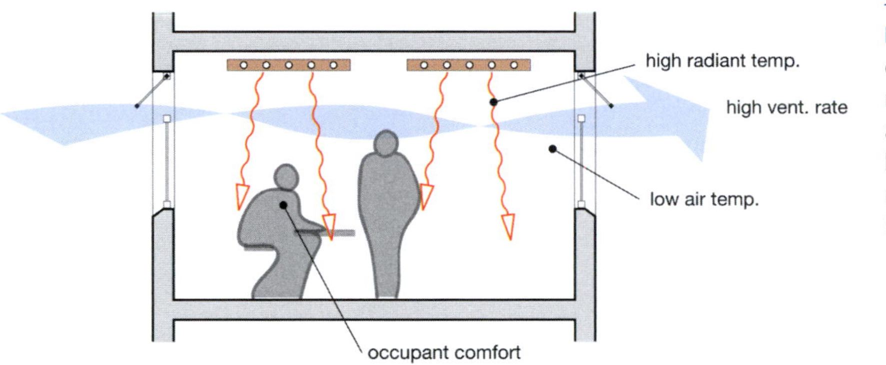

Figure 7.4 Providing comfort at a higher mean radiant temperature allows air temperature to be lower, thereby reducing ventilation heat loss

can achieve this, making them suitable for spaces such as workshops, sports halls and warehouses (Figure 7.4). Lowering the air temperature will lower conductive heat loss through the fabric, and loss by ventilation, which may be high from large volume spaces such as these. To some extent, underfloor heating also elevates mean radiant temperature, and delivers the heat at floor level, making it suitable for spaces with large ceiling height, including 'semi-heated' spaces such as atria.

High-temperature directly gas-fired heaters and electric incandescent heaters are available for high level installation in similar circumstances. Here, relatively small areas are heated to high temperatures (400°C to 850°C). The overall effect is similar to large areas of surface at a lower temperature. However electric heaters should be avoided if possible due to the large CO_2 emissions per kWh of electricity.

Positioning emitters

Radiators and convectors are often placed at the perimeter of rooms on external walls, and often beneath windows. Both can lead to high local air temperature, and if this is against a poorly insulated wall or a window, heat losses will be significantly increased. This is particularly true when windows are curtained, if the warm air is channelled into the space between the curtain and the window, as in Figure 7.5. Further issues arise during refurbishment programmes when cellular offices or meeting rooms are built around the perimeter of rooms with open plan areas in the centre. These refurbishments often make no provision for changing the heating system, resulting in unbalanced heated environments in all areas.

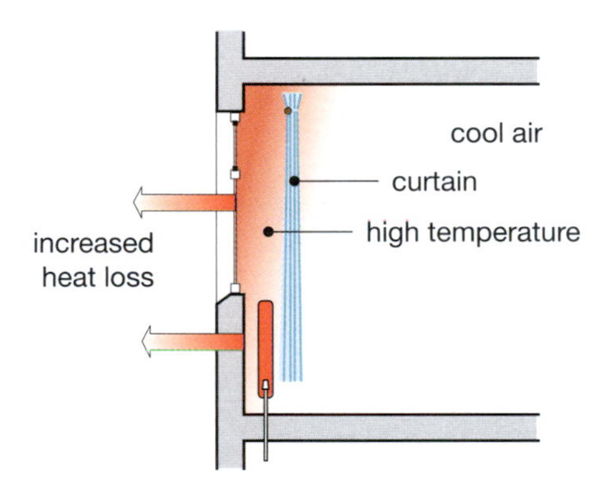

Figure 7.5 Poor positioning of heat emitter leads to increased losses due to high local air temperatures

Sizing emitters

Reduced heating loads give the opportunity of reducing emitter size. This may be an advantage in terms of space use, and a saving in capital cost. One consideration, however, is that the larger the output of the emitter the lower the water temperature needs to be for the same heat output. Lowering water temperature reduces non-useful heat losses from pipework, reduces flue temperature, and increases heat recovery in condensing boilers. Low water temperature also leads to higher efficiency when heat is supplied by heat pumps, geothermal and waste heat sources.

Underfloor heating is slightly different, since the maximum water temperature has to be low for comfort reasons (avoiding hot feet). The penalty is that a large area is required. This means that underfloor heating is a good choice for low energy use, provided the response time is appropriate for the heating regime (see Section 2.1).

Coolth emitters

Cool air is heavier than warm air, resulting in the natural convection currents illustrated in Figure 7.6 being reversed. Horizontal cooled surfaces behave differently from heated surfaces in that a cooled ceiling can deliver cool air to the floor, by downward flowing plumes, and a cooled floor creates a layer of stable cool heavy air. It is unusual to use conventional radiators or passive convectors for cooling purposes but in principle this can be done. All other emitters are suitable, thus it is feasible to use the same emitters for both heating and cooling. Cooled floors and ceilings carry much the same advantages as for heating applications, including their large surface area allowing small temperature differences. This may permit using ambient coolth sources such as ground or seawater.

However, a new problem emerges with coolth delivery – condensation. If the surface temperature is below the dewpoint, condensation will form. This is unacceptable for floors, and probably ceiling panels, but can be handled in fan coil units, where the collection of condensation effects a dehumidifying function.

7.4 Fans and pumps

CO_2 emissions due to the electrical energy consumed for circulating air by fans and water by pumps can be as large or larger than that for heating or refrigeration (Figure 7.1). Much of this is due to wasteful control – that is, moving more air or water than necessary – and this can be taken care of by improved controls, see section 7.7. However, the technical characteristics of the fans and pumps themselves affect energy use. Refurbishment may provide the opportunity to change to more efficient equipment in itself, or incorporate new features such as variable speed drives which permit a more efficient control strategy on ventilation air.

Ventilation fans will be designed to satisfy the highest annual demand and lower ventilation rates will be adequate to maintain air quality for much of the time. Variable speed drives can be retrofitted to fans and controlled by CO_2 levels in the exhaust air or by set profiles according to time of day.

Water circulation pump speeds can be controlled by pipe pressure so that where flow is reduced as valves close down due to thermostats sensing adequate temperatures, pump speed is reduced, greatly reducing power usage. Modern

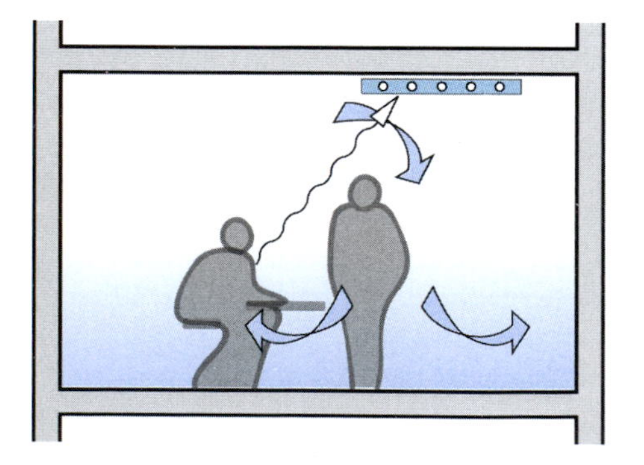

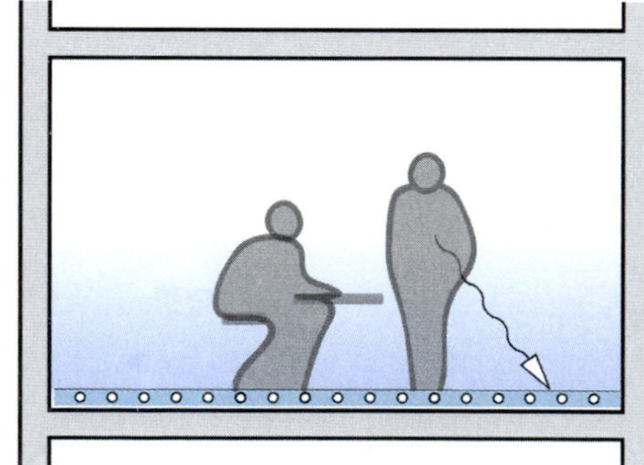

Figure 7.6 Downward cool air plumes from cooled ceiling panels (a) and stable stratified cool air from cooled floor (b)

pumps and fans are generally much more efficient than older equipment.

7.5 Refrigeration

Where active mechanical cooling is required in a building, this can be provided by a variety of systems giving varying efficiencies. These are mechanical central chillers, using one of:

- absorption chillers; or
- local 'split-unit' heat pumps.

Conventional refrigeration plant produces chilled water that circulates around the building supplying fan coil units or other emitters where necessary. Efficiencies vary between machines using different types of compressors.

Chillers can 'dump' heat from the building to the external environment via:

- wet cooling towers;
- dry cooling towers;
- using the ground.

Wet cooling towers give the highest efficiency but if not fully maintained have been linked as sources of Legionnaire's disease and as such require regular, effective maintenance and control. Fan-driven dry systems are becoming more common as they do not have any links to Legionella, due to the removal of moist, warm breeding grounds, but are a less energy efficient method of dumping heat.

Ground sourced systems use a horizontal grid of pipes at least a metre below ground, or a series of vertical pipes in which a fluid, normally water, is circulated and cooled by the ground. Ground pipes can give good chiller efficiencies, depending on ground temperatures and soil types, and the cooled water can be used through a heat exchanger directly to provide 'free cooling' for the building at modest demand levels.

In smaller buildings and even for supplementary comfort cooling in larger buildings, individual air cooled units are quite commonly used directly through walls, windows or ceilings. These normally provide cool air directly to the space with heat dumped via a fan coil to outside. Low efficiencies and poor control may make these installations inefficient for long-term use.

Where there is an available source of 'waste' high grade heat, in or adjacent to the building, absorption chillers can be used to provide cooling. These are similar to conventional chillers but use heat rather than electricity to drive the process of producing cold water. Absorption chillers are most commonly used where there is a combined heat and power (CHP) system close by, with surplus heat, particularly in summer. Overall efficiencies of absorption chillers are low and careful calculations are needed to ensure that they can in practice provide a low CO_2 cooling system. Absorption chillers have also been used in conjunction with solar heat sources.

Chillers are rated in terms of efficiency by their Coefficient of Performance (COP), which is the extracted heat energy divided by the electrical energy input. Instantaneous COPs can vary up to 4 or 5 for a high performance and/or magnetic bearing chiller, but it is the annual performance, taking into account periods of low demand, that is the key to low energy use. It must also be borne in mind that conventional electrical energy has a much higher CO_2 content than thermal energy.

7.6 Lighting installations

Luminous efficacy

For non-domestic buildings, lighting energy forms a major component of the total energy consumption, as shown in Figure 7.1. The adoption of a strategy to maximize the use of daylight in order to minimize artificial lighting energy has been discussed already in section 1.6. It involves maximizing the ingress of useful daylight by architectural means, and the use of controls to ensure that the artificial lighting is displaced by the daylight. The energy use is also dependent on the efficiency of the artificial lighting system itself, the subject of this section.

Luminous efficacy is defined as the ratio of luminous energy (lumens) to the power consumed by the source. The global luminous efficacy of a system is given by:

$$E_g = E_s \times UF$$

where E_s is the efficacy of the source and UF is the utilization factor. The utilization factor is the ratio of useful light on the workplane to that emitted from the source, and is a property of both the luminaire and the room.

The two factors on the right are both candidates for improvement. Table 7.1 gives generic descriptions of sources, luminaires and room characteristics to indicate the potential reduction of lighting power. The energy use is compared with a base case which might reasonably be regarded as a typical unimproved installation.

Table 7.1 Global lighting efficacy for various combinations of lamp, luminaire and room type

Lighting scenario	Lamp efficacy l/W	Global efficacy l/W	Installed power W/m² for 300 lux
Tungsten pendent diffusing globes surface reflectances med. maintenance poor	14	3.25	92
T12 38mm fluorescent magnetic (choke ballast) surface mounted battens with plastic diffusers (degraded) surface reflectance med. maintenance poor	55	16.25	18
tungsten halogen uplighters timber ceiling other surface reflectance med. maintenance average	18	3.45	87
quartz halogen recessed downlighters surface reflectances med. maintenance average	24	11.5	26
metal halide uplighters ceiling nom. white other reflectances light maintenance good	90	37.5	8
compact fluorescent triphosphor elec. ballast recessed downlighters reflectances med., maintenance good	64	32.0	10
fluorescent T8 tubes elec. ballast surface batten acrylic diffusers maintenance med.	90	42.0	7
fluorescent T5 tubes elec. ballast surface batten low brightness maintenance good	105	50	6

Note: Indicative values only.
Assumptions: Med. reflectance – ceiling 70%, walls 50%, floor 20%.
Utilization factor 0.65–0.75 downlighters, 0.55 uplighters white ceiling.
Light output ratios 0.5–0.85.

Illuminance level and distribution

As a generalization lighting levels tend to be higher than necessary. Table 7.2 gives recommended values. Replacement at the same power rating, with more efficient lamps and luminaires will only increase illuminance levels, not save energy. Thus the replacement specification must take account of increased light output, and possibly, reduce design illuminance to recommended values.

Table 7.2 Selected artificial illuminance levels non-residential buildings

Building type	Location	Service illuminance
General buildings	Corridors	100
	Stairs	150
	Reception areas	150
	Offices general	400
	Offices + VDUs	300
	Offices + task[1] lighting	150
	Drawing offices	700
	Drawing offices + VDU	500
	Conference rooms	300
Airports & transport buildings	Reception areas	250
	Circulation areas	150
Libraries	Bookstacks	150
	Reading areas	300
Schools and colleges	Classrooms	300
	Lecture theatres	200
	Laboratories	400
	Sports halls	500

Note: 1 Desk task lighting should provide average 700 lux over 1m^2. Values indicative only – may be covered by specific codes and regulations.

Recommended illuminance levels are based on the most exacting task to be carried out in the space. Large savings can be made by using task lighting, providing high (but controllable) illuminance over limited areas, and providing background illumination to the whole room at a much lower level.

For example, a 1000m^2 office has a uniform illuminance of 500 lux. If new high efficiency lamps and luminaires are used this could be provided at 10W/m^2, a total of 10kW. By reducing the ambient illumination to 150 lux, and providing a 12W task light to each of the 100 workstations (giving 600 lux over an area of 1m^2), the total load is reduced to 4.1kW. It is estimated that this measure alone would reduce annual CO_2 emissions by more than 2 tonnes. Apart from the direct saving of energy, this carries the further advantage of reducing unwanted internal gains during the warm season, thereby reducing overheating, or cooling energy.

This approach is particularly applicable to the modern office with LCD display screens, which are themselves luminous. If dimmable, task lighting allows the user to balance the brightness of illuminated documents approximately to that of the screen, or to their preference, improving visual comfort, providing adaptive opportunity, and possibly making further energy savings.

7.7 Controls

Local control

Any unoccupied part of a building that is being lit, ventilated by mechanical means, heated or cooled, is a potential waste of energy. In many buildings, workstation occupancy levels are often as low as 50 per cent, whilst there are many other areas such as stores, meeting rooms, rest areas and service areas that are occupied far less. A vital role of system controls is (or should be) to try to match environmental conditioning with occupation. This matching takes place over time and space and thus becomes the issue of time programming and zoning. A third issue is the degree to which systems can respond to unprogrammed occupation, for example by occupancy detection or user control.

The degree to which environmental conditioning can match occupancy is dependent on the speed of response of the system. Artificial lighting can respond instantaneously, and thus offers good potential for demand matching. For a single well defined room (such as a classroom or meeting room) this is completely possible, and automatic switch-off with manual switch-on is generally tolerated. In more diffusely occupied spaces such as open plan offices, and in particular circulation areas, it is more difficult. Walking towards an unlit area, that is only illuminated when occupied, would not generally be acceptable.

Heating (and cooling) does not respond instantaneously. This makes it more difficult for heating to respond to occupancy. However, some heating systems are quite rapid response, and if used in conjunction with a set-back – that is, a lower set temperature for standby – this form of control can be quite acceptable. For example, if a room is unoccupied for more than ten minutes, the set temperature drops 4°C.

Mechanical ventilation consumes energy both from the fan power and the resulting heat loss. There is no virtue in changing air that already meets accepted air quality standards. Local demand control can be achieved quite easily using CO_2 detectors. This is preferable to occupancy detectors since it will respond to the numbers of occupants in the space, rather than just whether the space is occupied or not.

Central control

The key issues here are time programming and zoning. Time programming is technologically simple, and with digital devices, the cost is becoming trivial. The problem with time programming is whether the management of the system allows the flexibility to alter the programme to respond to changing circumstances.

For example, a meeting room is used three times a week, usually in the afternoon, but occasionally (once per month) in the morning. Does the facilities manager tell the energy manager to programme for every day, every afternoon, or respond to each meeting? We can probably anticipate the answer! The difference between the three options would be:

All the time	100%
Afternoons +	52%
Occupation only	30%

Thus up to 70 per cent of the energy used by that room is wasted. Clearly a room booking system integrated with the heating programme could solve the problem, although it still could not cope with unpredicted spontaneous use.

However, even spontaneous meetings do not occur with zero lead time – participants have to be informed, documents gathered, coffee and biscuits arranged. If on checking by the intranet that the meeting room is free, the originator of the meeting could call environmental services for that room to be activated, by the time the meeting was under way it could be up to acceptable comfort levels.

The actual spatial and temporal occupation density of all buildings is very low. Studies have shown that even for a nominally *fully* occupied office, the occupation of workstations is actually about 50 per cent. Other areas are even more sparsely populated. Different building types such as schools, hospitals, hotels and even residential buildings, show a similar characteristic.

Technical solutions seem to lie jointly in the area of communicating with the central programming system, and responding locally by passive occupancy detection or by user control. The latter is, of course, what we have been doing for centuries, before the concept of central control existed, and thus before the idea that an unoccupied space should be held permanently 'at the ready'.

Zoning

Occupancy demand matching can be achieved only if the centralized services are zoned appropriately. Existing buildings have often had additions to both fabric and services, and changes of use, which have led to the original zones becoming redundant.

For example, a community centre was built with a small hall and foyer, on a single heating circuit. Later a large sports hall is added, with a second heating circuit and given independent control. However both halls shared the use of the foyer (including toilets), and the management had to heat the whole building when the large hall is used. Quite a small modification to the pipework, and the addition of two motorized valves, now allows any combination of the three spaces to be programmed.

The concept of zones applies to centralized control. As zone size decreases, down to individual rooms or even individual appliances (e.g. heat emitters) the issue then becomes between central control and local control. A room thermostat is an automatic local control whilst an on/off switch is a manual local control. If all control were at this level, it might appear that zones would become redundant.

It is interesting to note that, traditionally, the zoning concept is to exert control at a local level, by communicating via the actual service being provided – electricity, heated water or conditioned air.

Developments in IT have made exciting new possibilities. The ability to address an actuator – a valve, a damper, a light switch – using a digital signal recognized only by that specific component, has been available for some time, although it seems that its application is not widespread. This has been partly due to cost – low-volume production, and also the cost of hard wiring the data bus to the component.

But we only have to look to other application areas to see the cost of *wireless* digital control reducing by orders of magnitude, and we can envisage a day when every component will incorporate its own digital receiver. Environmental components simply have to be connected to a supply – water, air or electricity – and a receiver/actuator unit placed with them. Sensors could also be part of the same communication network, acting as control points in optimal positions.

Virtual zones are then created, simply by grouping together component addresses. This would have massive advantages in flexibility, and with appropriate software, building energy managers could configure complex zones. Conditional actions could easily be programmed in – for example – if the window sensor detects *wide open*, then *heating off*, or if *blinds down and no occupants* then *lighting off*. Both of these examples show the way the controls interact with individual's actions in a way that moves the result towards a low energy outcome.

The system could easily interact with a bookings diary for intermittently used spaces, ensuring that there was good occupancy demand matching.

7.8 Lighting controls

Lighting controls have to prevent demand mismatch in two ways. Firstly, to avoid artificially illuminating unoccupied spaces, and secondly to avoid artificial illumination when there is sufficient daylight.

Occupancy detection

The former can be achieved using centralized programming but this is unusual. The commonest method is to use occupancy detection in a room or zone. Artificial lighting in a room or zone is extinguished after a certain period of non-detection of the presence of an occupant. The detectors use passive infrared (PIR) or microwaves and detect movement. This has the slight problem that if the occupant is very still the detector may falsely detect non-occupation and switch off the lighting. Most commonly in areas such as toilets, stores and circulation spaces, the control will provide switch-on and switch-off, whereas in areas such as meeting rooms, offices and classrooms the control operates switch-off only, relying on manual switch-on.

Although the control is local, zoning is required because the occupancy detector will have to switch a number of light fittings. These must be on a common and exclusive circuit unless luminaries are installed that contain individual

Table 7.3 Selected daylight factors

Building type	Location	Average DF%	Minimum DF%
General building	Reception areas	2	0.6
	General offices	3	1.0
	Drawing offices	5	2.5
Airports & transport buildings	Reception areas	2	0.6
	Circulation areas	2	0.6
Libraries	Bookstacks	3	1.0
	Reading areas	6	1.5
Museums and art galleries	General	5	1
Schools and colleges	Classrooms	5	2
	Lecture theatres	1	0.3
	Laboratories	5	2
	Sports halls	5	3.5

Note: These recommended values originated from mid Europe. For the same yield of useful daylight, values can be reduced by 40% for latitudes less than 45°N and increased by 20% for latitudes above 55°N.
For more detailed version of this table please see CIBSE: *Window Design Applications Manual*

occupancy/illuminance detection exclusively for that luminaire's control.

Stand alone occupancy-detecting light controls, are low cost (typically €30–€100) and readily available. In buildings with intermittent room-use large savings can be made and the installation of these controls during refurbishment should be high priority.

Daylight detection

Most non-domestic buildings, are technically day-lit – that is, there is at least a 2 per cent daylight factor[2] (DF) for 80 per cent of the floor plan – but they spend most of their occupied time lit artificially. This waste is partly due to the lack of feedback to the occupant – that is, although the occupant notices under-illumination, there is no discomfort feed back from over-illumination. Thus, unlike closing a window once the room has cooled down, turning off the electric light when there is sufficient daylight becomes an ethically or intellectually driven action and not discomfort driven.

Photo-sensitive detectors can be used to make up this deficiency. The most basic system monitors the illuminance in the room and if the illuminance value is above a datum (the design illuminance + the artificial illuminance), the artificial lighting will be switched off. These systems first appeared in the 1970s and had mixed reception, occupants often finding them irritating due to the sudden change of illuminance.

A whole series of refinements followed, the most important being dimming control, which allowed the artificial lighting to be gradually dimmed, as the daylight illuminance approached the design illuminance. This not only improved visual comfort, but increased the energy saving. Other refinements included the incorporation of a time lag, to avoid response to short term daylight fluctuations, and a system where the switch-off

2 DF = ratio of horizontal illumination in the building at the point of interest to the unobstructed illumination outside the building, expressed as a percentage. Recommended values are shown in Table 7.3.

action occurred at specific times of the day.

Systems installed in open plan offices were usually under full auto control, whereas systems in cellular offices and small rooms were often auto off and manual on. This created a further energy saving due to the phenomenon of increased tolerance of failing daylight, where people will delay switching on artificial light even when the daylight has reduced well below the accepted artificial lighting level.

All manual switching can have some energy saving potential. It is found that in cellular offices, with one or two occupants, it is more likely that the lights will be switched off when there is sufficient daylight than in open plan offices. This is probably due to territorial effects and the 'taking ownership' of control, and is an extension of what we would expect to see in a domestic situation. However, even at this scale, zonal switching is best, allowing the lighting in areas of the room closest to the window to be switched off first.

Zoning

Two methods can be adopted:

1. The available daylight can be measured outside the building (usually on the roof) and then daylight illuminances calculated (using measured daylight factors) for various switching or dimming zones in the building. This is a centralized system and would be part of the building energy management systems (BEMS).
2. The illuminance for each zone is measured at a representative point in the zone, and used to control that zone only. Since it is inconvenient to place the sensor on the workplane facing upwards, it is often placed on the ceiling facing downwards. If this is done, it must be recognized the sensor will not measure the actual working illuminance but the ceiling illuminance, which is affected by the floor and work-surface reflectance. The latter system is becoming more popular.

In both of these systems, light fittings must be grouped together into zones, controlled by a single detector or predicted illuminance point, and these zones must relate logically to the variation of daylight in the space. That is, they should be approximately defined by isopleths of equal daylight factor (Figure 7.7). For most side-lit areas this simplifies to having two (or possibly three) banks of light fittings parallel to the window wall and at a distance of about 1.5 and 4.5 metres from the window wall. Zoning of this kind will show significant savings over a single zone for the whole day-lit area.

Energy savings

The savings created by photo-sensitive control are considerable. Figure 7.8 illustrates the effect for both on/off control and dimming control. The low cost of the control system makes them highly cost effective and their incorporation absolutely essential in any non-domestic building refurbishment.

7.9 Building energy management systems (BEMS)

BEMS are now used in virtually all mechanically conditioned buildings and increasingly in naturally ventilated buildings. Potentially they have the ability to control energy use and record energy consumption, as well as controlling the internal environmental conditions.

BEMS can be used to control all aspects of the building – lighting, heating and cooling plant, internal temperatures and air quality, ventilation systems, night cooling, optimal start/stop and solar control blinds. Most systems have a visual display that can show the operator the current conditions and settings throughout the building and enable them to make modifications.

Although theoretically BEMS enable the efficient control of a building, in practice many systems do not give the good results. Poor system design, poor programming, over-complex systems and poor understanding by the operator

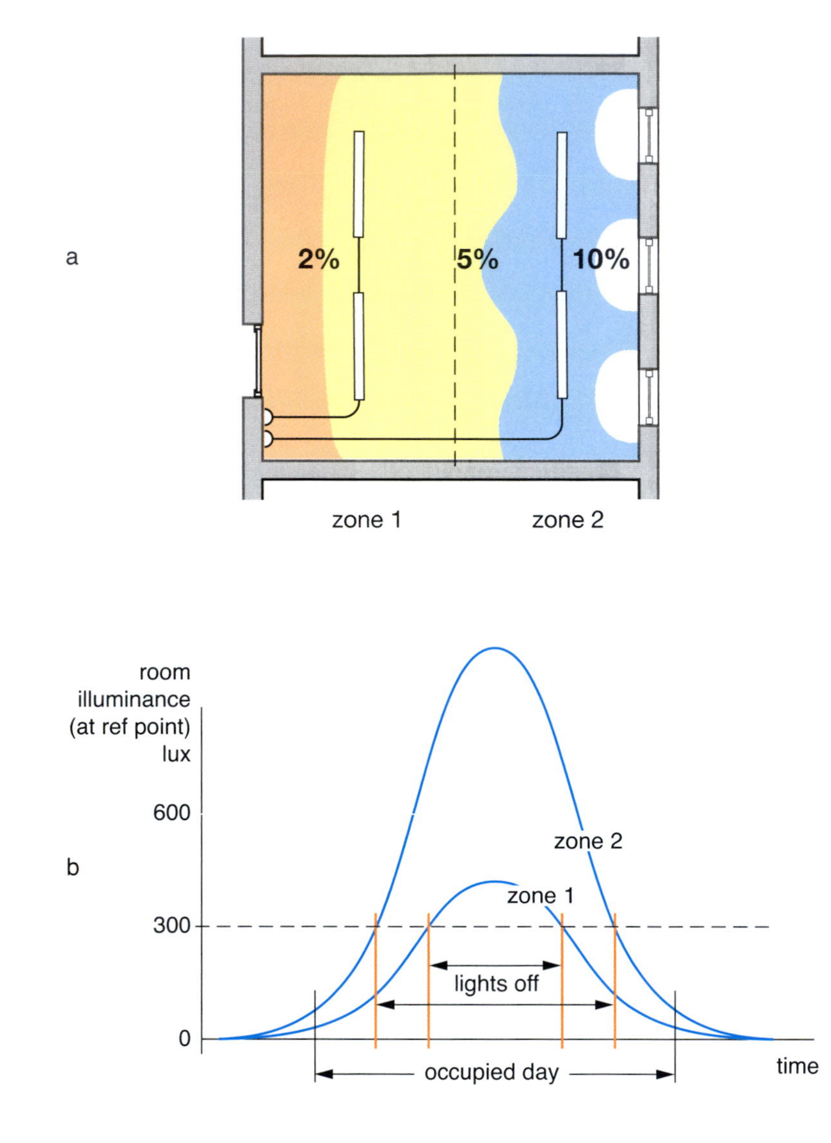

Figure 7.7 A daylight factor contour map of a side-lit room showing position of two separately switched banks of lighting (a) and the resulting switch-on times (b)

With the steady reduction of costs of micro-electronics, luminaries are now being manufactured with their own detector – that is, one per fitting. Potentially this offers very good control, but only provided the switching or dimming threshold is set up properly. Pre-set values will not recognize the effect of work-surface or floor reflectance.

can lead to systems being overridden or disabled. Although a single BEMS can control all energy and comfort aspects of a building, using either hard wiring or radio links, separate sub-systems are sometimes found controlling different physical or subject areas – for example, local controls for split-unit cooling in meeting rooms. An important design aspect for the BEMS contractor is to ensure that the level of complexity, and the design of the display and operating system, fit the likely level of knowledge and training of the person who will be expected to operate it on a day to day basis.

7.10 Adaptive controls

The principle of adaptive control has been described in some detail in section 1.5. Essentially it is the inclusion of the occupant in the control loop, that is, permitting the occupant to take part in the control process without allowing the action to destroy the automatic control strategy. The advantage of this approach is that, given a level of control, the occupant is much more tolerant of non-neutral conditions, as illustrated in Figure 7.9.

The inclusion of the occupant in the control means that opportunities must be provided – operable windows, blinds, heating/cooling controls, desk or ceiling fan controls.. These are technical or element-based opportunities. Other opportunities could be termed behaviour-based, such as a cool 'chill-out' rest area, cold-drinks machine, or relaxed dress code. These are also discussed in section 1.5.

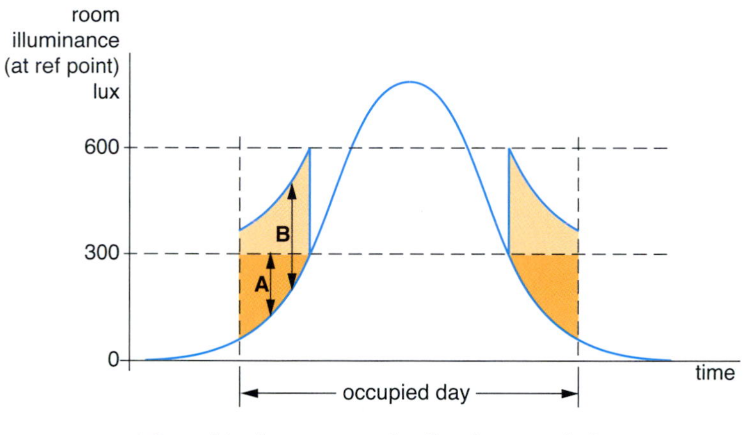

Figure 7.8 Room illuminance during the day for dimming control (A) and on/off control (B), indicating the improved energy saving for dimming control

Two issues dealt with here are *feedback* and *caretaker function*.

Feedback

Figure 7.10 shows the interaction of automatic control with a common action such as opening a window. The occupant feels warm, opens the window, cool air enters and the temperature drops, the thermostat calls for more heat, and the occupant forgets to close the window; the thermostat has broken the feedback loop – result, energy waste.

The conventional engineering response to this has been to say 'if the temperature was right in the first place, the occupant wouldn't have opened the window, so we control temperature even more closely and don't provide operable windows'. Note that this answer is dependent on the conventional (Fanger-based) principle that there is a 'right temperature'. However, this occupant might have just hurried to work, climbed two flights of stairs and be warmly dressed for the prevailing climate – resulting in a very slightly elevated core temperature. He/she now needs a period to 'cool off' – a reassurance to the unconscious that long-term heat balance can be restored.

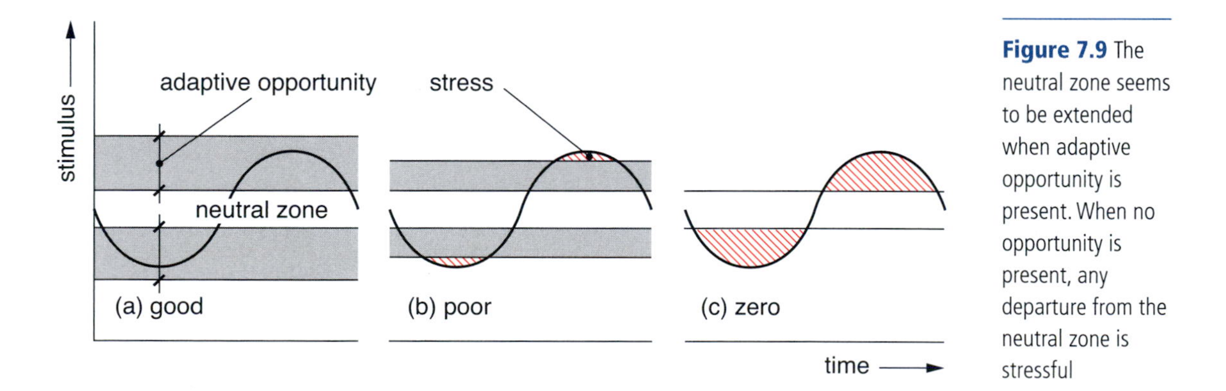

Figure 7.9 The neutral zone seems to be extended when adaptive opportunity is present. When no opportunity is present, any departure from the neutral zone is stressful

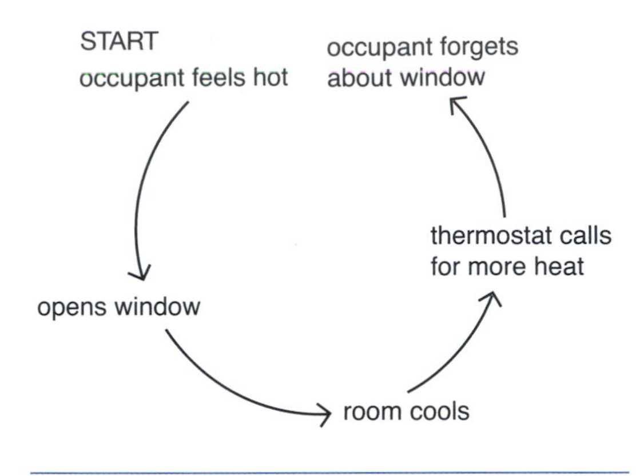

Figure 7.10 The effect of thermostatic control to break the natural feedback loop in occupant control

The correct engineering response should be to detect that the window is open and deactivate the local thermostat, or at least give it a temporary set-back. Only when the window is closed, would the original set-point be restored.

This system has been installed in the Revival GSIS building to control air-conditioning in the perimeter zones with operable windows.

Open plan situations create more of a challenge to achieve this. However, as we move towards the systems described above using digital addressing and virtual zoning, response such as this becomes increasing possible even in multi-occupied spaces.

Another case where controls can break the feedback loop is in photo-sensitive lighting control. Consider the case where a room is equipped with moveable and modulating louvre blinds. The occupant experiences glare, deploys the blind leaving its modulation at closed. Insufficient light is transmitted and the daylight detector switches on the artificial lighting, or in the case of manual switch-on, allows the occupant to do so. If this action were prevented by detecting that the blind was down, the occupant would be obliged to control the blind transmission (by adjusting the louvre angle) until there was sufficient light.

Caretaker controls

This is closely related to issue the above – that is, it is a function that prevents unnecessary waste but allows the occupant short excursions into what might be regarded as wasteful behaviour. The ability to override automatic control is an important feature of adaptive control. Caretaker control is all about gently returning to automatic control when it leads to lower energy consumption than the present status.

For example, an occupant wants a higher temperature. If he/she is lucky enough to have access to a thermostat, they will quite probably adjust it to a higher temperature than desired, under the false impression that it will achieve this condition quicker. It will then remain at that setting, until the occupant feels too hot, but quite possibly prompting him/her to remove some clothing or even open a window.

A caretaker control will only offer the occupant a warmer/cooler option about a long-term average value. It will then return to that setting over a time span of about 30–60 minutes. It will detect the long-term preferred setting from the occupant's 'votes' with the warmer/cooler button, and could even add the sophistication of gradually lowering this in the heating season and raising it in summer (if mechanically cooled). This will encourage the occupant to make seasonal adaptation to their dress ensemble.

7.11 Hybrid and mixed mode systems

This relates to mechanical cooling and air-conditioning and is the term used to describe the intermittent application of mechanical cooling, to prevent extreme excursions of high temperature. This permits a predominantly passive strategy to dominate the building's operation, leading to much lower energy consumption than if the building were conventionally air-conditioned all year round.

This approach is no different in principle to the heating of buildings – which is partial and intermittent, and as buildings reduce their heat

Figure 7.11 Use of backup chillers in Daneshill House in the mechanically driven night-ventilation system is not evident to the occupants

demand because of high insulation, will become increasingly less frequent.

Mixed mode is also used to describe a spatial mix – that is, where some spaces may be air-conditioned (e.g. deep non-perimeter areas, densely occupied areas such as lecture theatres, meeting rooms, etc.), whereas other perimeter zones may be naturally ventilated only.

In both cases, the issue is to make sure that the predominantly passive operation does not unduly compromise the performance when operating mechanically, and vice versa. This is largely achieved by controls, and relates closely to the issues discussed in adaptive control.

Other important considerations are to ensure that the changeover from passive to mechanical mode takes place at the appropriate point. Otherwise there is the risk that the building will become air-conditioned by default, possibly with an inadequate system.

It is also a good idea if the presence of comfort cooling can be disguised. This is to prevent occupant demands for mechanical cooling at times outside the target need periods, in the belief that it will always increase comfort. The ability to disguise will depend on the systems employed. For example cooled floor or chilled ceiling panels, can operate without obvious signs. If mechanical fresh air supply is already present, then tempering this air can also be achieved without advertising the fact.

Where it is impossible to disguise, such as the activation of a chilled air supply that is not normally in operation, then the feedback controls (as discussed in section 9.9) should be in place to prevent wasteful use, and to give the occupant some control over the decision to mechanically cool or not.

8 Renewable Energy Options

Refurbishment may provide the opportunity to incorporate renewable energy (RE) technology. Many applications of RE technology are relatively independent of building work, and could be retrofitted at any time – such as solar panels and wind turbines. These applications are adequately described in numerous resources. We concentrate here on instances where there may be savings to be made when the RE technology is applied as part of fabric or services improvements.

The cost effectiveness of RE technology ranges widely. For example, biomass (wood pellet) boilers are close to cost effectiveness, whereas photovoltaic (PV) applications, where grid connection is available, are a long way from cost effective. However, this does not negate the value of them since it is widely recognized that as conventional energy costs rise, and volume production reduces equipment costs, they will eventually become cost effective. Meantime it is essential that experience is gained with real applications. Furthermore, regulated obligations now exist in some countries where a fraction of on-site renewable is required for new developments, and it is quite likely that similar obligations will be extended to refurbishment projects in the future. For example, in the UK where the building or mechanical services are being extended, 10 per cent of energy demand from renewable resources should be provided where paybacks of less than seven years are achievable.

8.1 Other renewable energy technologies

Most other RE technologies do not interact directly with the building and thus refurbishment does not offer any particular constraints or opportunities. Solar thermal and PV applications on angled frames on flat roofs, and on-site wind turbines, fall into this category.

Table 8.1 overleaf

Table 8.1 Application of various RE technologies and their interactions with building refurbishment

Renewable energy technology	Application	Comments
Solar thermal (evac. tube)	Upgrading domestic hot water system for hotels, hospitals, schools	Installation may be integrated with roof refurbishment – close to cost effective
Solar thermal (flat plate)	Low temperature applications such as swimming pools	Even unglazed flat-plate collector can be effective for this low-temperature application
Photovoltaic	Re-cladding panels and roof tiles	
Photovoltaic/thermal	Re-cladding with air-cooled PV panels.	Electricity generation and ventilation pre-heating. Cooled panels work at higher efficiency
Photovoltaic	Opaque PV used as shading devices	Geometry for optimum collection and shading tends to coincide
Photovoltaic	Semi-transparent PV used for reduced transmission glazing panels in large spaces such as atria	Not optimum shading since PV is about 85% absorber and re-radiates absorbed energy inwards
Biomass heating	Biomass heating requires space for fuel delivery and storage	Local emissions regulations need consulting
Ground source heating	Uses a heat pump	Operates at low temperature requiring an appropriate delivery system, normally underfloor heating
Ground source cooling	Uses a heat pump	Increases efficiency of refrigeration due to lower temperature cold sink. Displaces electricity. Often used in conjunction with heating
Solar thermal (cladding collector)	Re-cladding in conjunction with external insulation	Heated air collected between lightweight absorber and external insulation – best for ventilation pre-heating
Solar thermal (evac. tube)	Contributing to space heating in buildings with low heat demand and integrated storage system	Installation may be integrated with roof refurbishment – not yet cost effective

Note: A number of innovative techniques (examples are the last two in the table) have been attempted from time to time but as yet are unlikely to prove cost effective.

Part Three **Case Studies**

9 The Albatros, Den Helder, The Netherlands

Double skin applied to residential tower from the 1970s in sustainable conversion to offices

'The Albatros' is a high-rise block of seven floors ($4400m^2$) with a ground floor of meeting rooms, reception and circulation ($1800m^2$). The original building, built in 1972, had the major glazed facades orientated east–west, single glazing in aluminium frames, no wall insulation, inefficient artificial lighting, heating by radiators and gas boilers, and a mechanical ventilation system without heat recovery. The building was originally used as housing for officers of the Royal Dutch Navy and had very high energy consumption.

Objectives

- 30 per cent reduction in energy demand compared with current newbuild regulations.
- 30 per cent energy supply from renewable sources or a 50 per cent reduction in overall energy use.
- Any over-cost of eco-refurbishment to be minimized.
- Internal conditions satisfying all newbuild criteria.

Refurbishment strategy

The application of insulation to the original building, either inside or out, would have resulted in many thermal bridges, for example where internal partitions meet the external wall and where the balcony meets the external wall. This suggested that the insulation of the facade should be based on a second skin applied over the existing skin and supported by the existing balcony structure. The change of use to offices, with increased occupancy density compared with residential use, underlined the need for good daylighting and energy efficient ventilation.

Figure 9.1 The Albatros before (left) and after refurbishment

Table 9.1 Different refurbishment specifications

	Consolidate	Upgrade	Second skin facade	New	Preliminary design second skin facade
glazing	glazing panels in front of existing facade	new windows with low-e glazing	new wooden frames with low-e glazing	low-e glazing in wooden frames	existing facade with new openable windows
insulation		insulation of opaque envelope	insulation of opaque envelope	insulation of opaque envelope	insulation of opaque envelope
airtightness			improve airtightness of facade	improve airtightness of facade	improve airtightness of facade
ventilation		high efficiency heat recovery	nat. supply self-balancing vents, mech exhaust	nat. supply self-balancing vents, mech exhaust	nat. supply self-balancing vents, mech exhaust
heating		condensing gas boiler	heat pump and condensing gas boilers	heat pump and condensing gas boilers	heat pump and condensing gas boilers
lighting		high efficiency lighting	high efficiency lighting	high efficiency lighting	high efficiency lighting
interior walls		existing	new	all new construction	new

Note: 'New' includes demolition and new structure.

An important strategic issue was that options, shown in Table 9.1, should be selected on the grounds of life-cycle performance; that is, total embodied and consumed CO_2. However, contrary to normal practice, a single (long) assumed building lifetime was not used; rather, options were considered at various future dates to see how savings evolved. This is illustrated in Figure 9.2. The performance as indicated by total CO_2 emissions (embodied + service) as a percentage of the existing building – for example, the 40-year bar for the 'preliminary design' indicates a reduction of 60 per cent compared with the 40-year total for the existing building. Note that all measures in 'consolidate' and 'upgrade' more than recover the invested CO_2 within five years, whereas the 'new' requires more than ten years. The 'preliminary design' took the most promising features from the other scenarios.

A summary of the final design choice is as follows:

- new glazing and aluminium frames with U-value 2.2W/m^2°K;
- second skin with sun shading in cavity;
- insulation of the existing opaque envelope with U-value 0.3–0.4W/m^2°K;
- passive ventilation system with wind pressure – independent vents;
- district heating with waste heat from on-site power station;
- heating with convectors in offices;
- lighting capacity 12W/m^2 with photo sensitive dimming control.

The double skin

The second skin is hung in aluminium framing

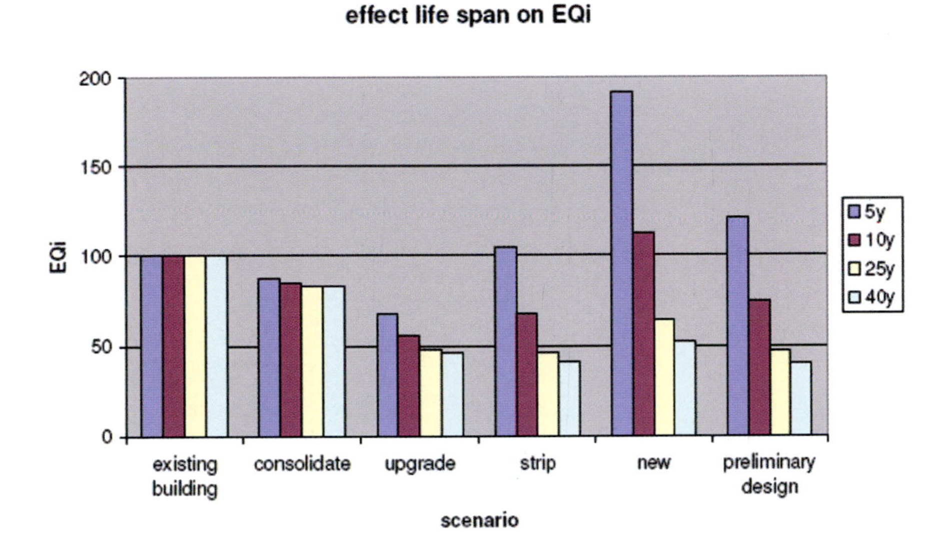

Figure 9.2 Total embodied and service CO_2 (as percentage of existing building) for different refurbishment specifications, and for time spans of 5, 10, 25 and 40 years

onto the existing steel balcony structure. The steel structure was first removed to allow the cold bridges to be insulated. The existing single glazing is replaced with low-e double glazing in new thermal-break, openable aluminium frames.

The active second skin consists of 66 per cent of toughened single glazing and 34 per cent ventilation grill controlled by the building energy management systems (BEMS). In winter, the large outer vents are almost closed, and self-balancing ventilation[1] inlets in the inner skin (original envelope) allow pre-heated ventilation air to be drawn into the building. Air is extracted mechanically via extract grilles in the ceiling and on to a heat recovery unit and heat pump. The section is shown in Figure 9.4 and a simplified elevation in Figure 9.5. Note that, moving horizontally, the grills alternate the function of inlet and outlet, minimizing the risk of short-circuiting.

Figure 9.3 The second skin. The walkway grid is at cill level; there is airflow from the inlet below the grid, to an outlet above the ceiling grid

1 As the pressure difference across the vent increases, e.g. due to wind, the aperture closes up resulting in an almost wind-independent flow rate.

In summer, the large inlet and outlet grilles in the second skin are open to allow high ventilation rates of the void, to keep temperature increment above ambient to a minimum. Extra vents in the inner skin open under BEMS control to allow night ventilation, driven by extra fans extracting from the corridor outside the offices. When the solar irradiation is greater than 200W/m^2, louvered blinds with a total transmission of 20 per cent are deployed in the double skin void on the inside of the outer glazing.

Performance of double skin

The thermal performance of the double skin in winter relies upon the increased temperature in the void compared with the outside temperature. This leads to reduced heat loss across the original facade, and reduced ventilation heat loss also, although drawing in cold air to replace that taken by ventilation will reduce the temperature in the void. The temperature in the void is elevated for two reasons – firstly heat loss from the heated building, and secondly from solar gains when they are present. Thus the double skin system has both a heat recovery function (sometimes referred to as dynamic insulation) and solar collection.

In order to investigate the potential contribution in this mode, temperatures were monitored in the room, the void and outside, and are shown in Figure 9.6. Temperature in the void is seen to track the outside temperature quite closely (except on sunny days), with an average increment of less than 3°C. In seeking an explanation for this disappointing performance, it was discovered that the large summer ventilation grills were not closing properly and the void was being ventilated at much higher rates than necessary. The BEMS was also opening the vents in the daytime, as late in the year as October. Both of these errors have now been corrected.

Ventilation and heating

Ventilation has already been partly described in connection with the double skin. After the air is

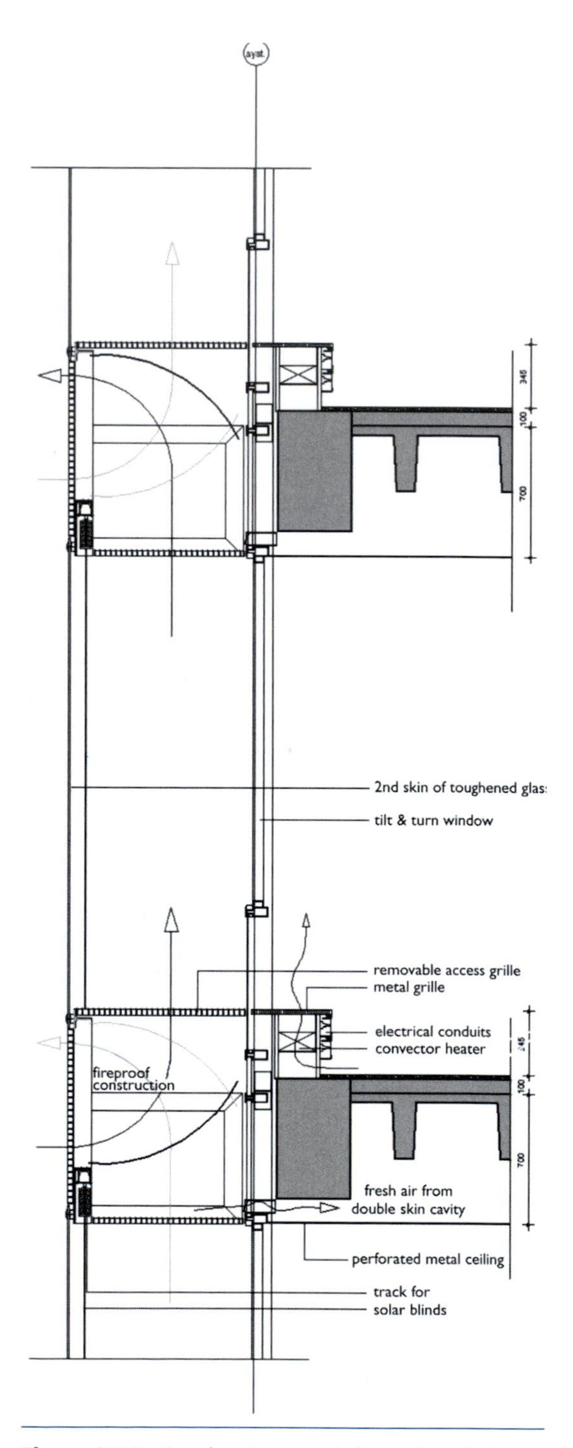

Figure 9.4 Section showing second skin and ventilation openings

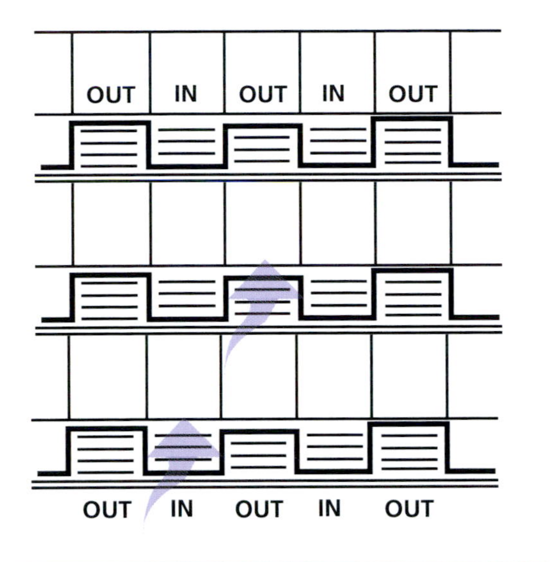

Figure 9.5 Simplified elevation showing alternate inlets and outlets

pre-heated in the double skin, it enters the office room at high level above the head of the window. Being cooler than room air, it descends in a turbulent plume and mixes with the warmer room air. This vent can be operated by the occupant, but is closed automatically at night in winter. Air is then extracted from the room by mechanical exhaust. Heating is provided by a natural convector installed at skirting level on the outer wall (see Figure 9.4). The emitter is generously sized, allowing a water temperature of only 55°C.

Performance

Indoor air quality was assessed by CO_2 measurement and it was shown that the hybrid ventilation scheme (mechanical extract, passive supply) was very effective. Figure 9.7 shows a typical trace for a 50-day period, during winter 2007, for one of the west facing offices. The daily occupancy pattern is clearly visible, and it is evident that peak concentrations are never higher than 900ppm, whilst average during the occupied period is about 650ppm. These are very good compared with 1200ppm (the generally permitted level), and also compare well with the design target of 1000ppm. However, the occu-

Figure 9.6 Temperature measured in the office, the double skin, and outside for period 18–27 October 2007

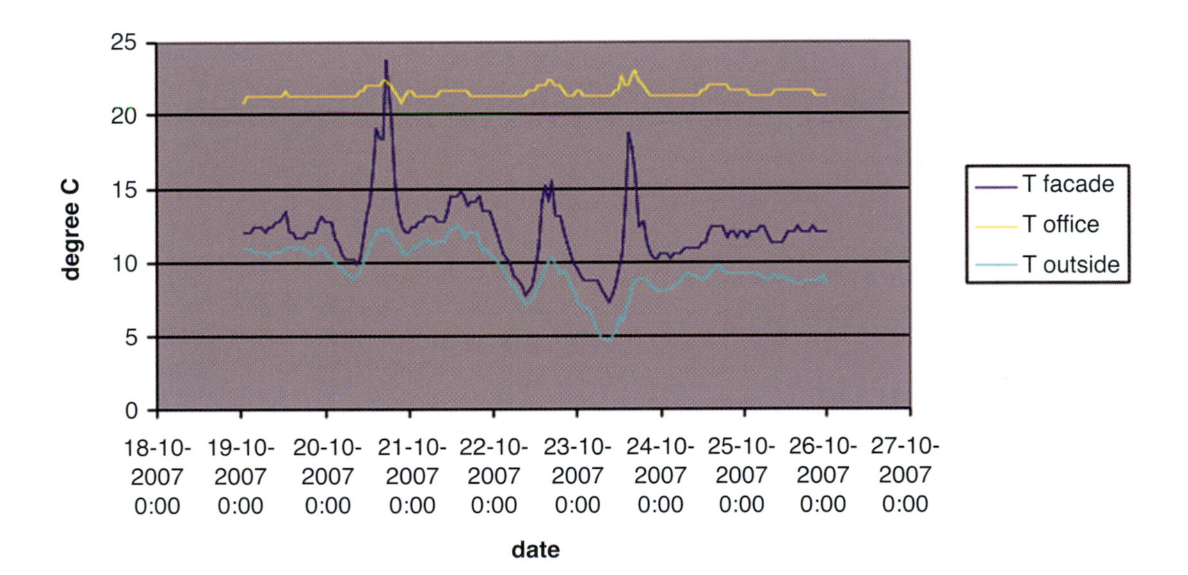

pancy density is low, with a global space provision of 28m²/person.

There was some concern that on the occasion of strong easterly or westerly winds, there might be cross-contamination between offices, leading to higher concentrations on the downwind side. This was checked with tracer gases and it was found that the movement of air between the zones is no more than 10 per cent of the fresh air flow into each room. These measurements also showed that the overall ventilation rate was around 0.65ac/h. This is quite low, but the fact that there is very satisfactory air quality proves that with good spatial efficiency (i.e. ventilation where it's needed) low ventilation rates are acceptable – leading to low ventilation heat loss and fan power.

Daylighting

Lighting energy typically forms a large proportion of total energy use for offices so a strategy for maximum use of daylight was adopted in the Albatros refurbishment. The potential was good, since the cellular offices are only five metres deep and the existing glazing area was large (more than 65 per cent of facade). However, the addition of the second skin, together with shading devices and the existing balconies, could cause significant obstruction.

The design solution adopted was to keep the full width, cill to ceiling glazed area, and to specify high reflectance finishes to all room surfaces and all surfaces in the second skin void.

Figure 9.8 shows the resulting daylight factor (DF). Two important results are evident; firstly the DF does not drop below 2.5 per cent, and secondly that the rise to approximately 14 per cent at the window is modest for a side-lit room, and is gradual. These characteristics are positive attributes for both daylight quantity and quality. They are achieved partly due to the high internal reflected component from the highly reflective surfaces, and partly due to the overhanging and reflective structures (including the shading louvres) located in the second skin void. These structures have the effect of redirecting light onto the ceiling and side walls, and thence by reflection to the back of the room.

The artificial lighting installation is of high efficiency high frequency sources and luminaires. The installed capacity is high at 12W/m², with a design illuminance of 500 lux. There is photosensitive dimming control to two banks of lighting at 1.5 and 4.5 metres away from the window. There is no occupancy detector, but there is an automatic switch-off at 1230 and 1800 hours. Switch-on is manual.

Overall energy performance

The energy use of the building, heat and electricity, was monitored for the period October 2006 to October 2007. Table 9.2 compares the

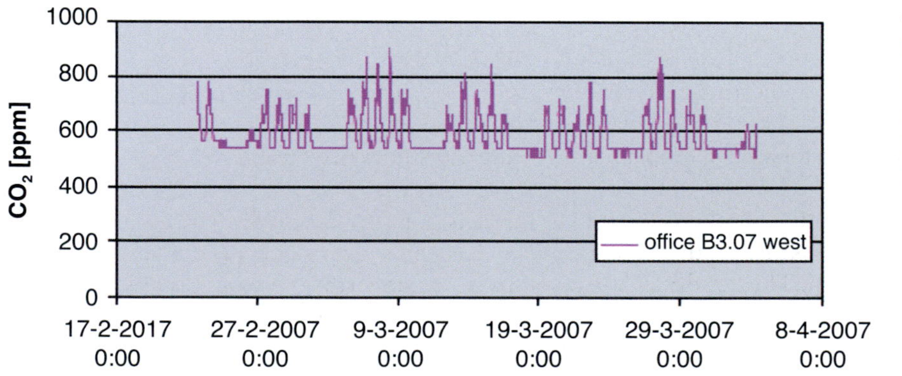

Figure 9.7 Measured CO_2 concentration for office B3.07 west, for 50-day period in winter 2007

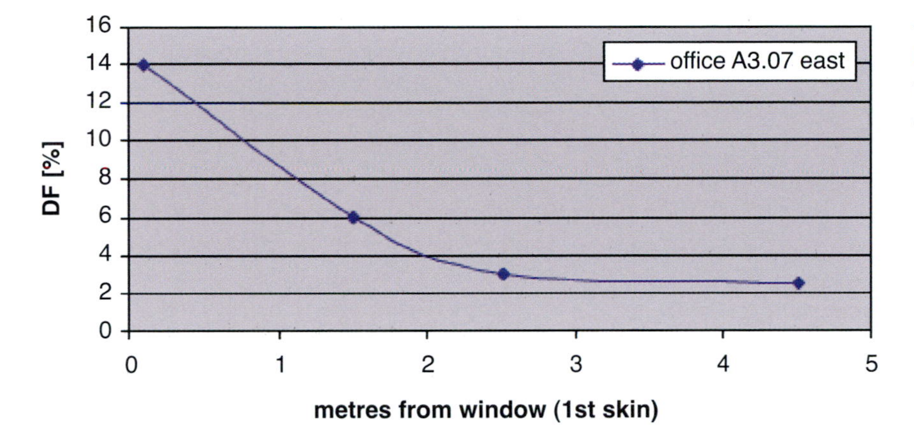

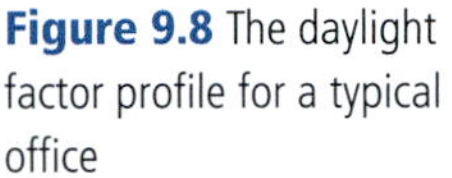

Figure 9.8 The daylight factor profile for a typical office

measured consumption with the original calculations, with a new building built to current regulations, and an average office building for the Netherlands. The measured heat has been corrected for the actual degree days, to the long-term value used in the predictions.

Table 9.2 Energy use in delivered kWh, October 2006 to October 2007

	Heat	Fans	Lighting
Albatros predicted	65	6	15
Albatros measured	154	2.2	3.6
Albatros corrected for standard occupancy	86	1.2	2.0
new office building	80	8	20
average office building	150	12	30

The first impression is that the heating performance is disappointing. However, three main explanations have been put forward. Complaints of cold discomfort from the inlet vents in the offices led to an inspection of the closure of the large vents in the second skin and they were found to be partially open. The resulting low temperatures, measured directly, as shown in Figure 9.6, were the result of an overlarge ventilation rate to the second skin void.

Secondly, the actual internal gains from equipment in the building were much lower than that assumed in the calculations. Thirdly, it was found that the BEMS was calling for heat for 24 hours per day, 7 days per week. This had been set up to cover out of hours use; in future this will be achieved by a special override switch.

Correcting for a conventional office occupancy profile reduces the annual heat demand by about 44 per cent which brings it into the region of a new office building. It is anticipated that the correction of the technical fault with the closure of the ventilators will reduce this further.

The electrical consumption (Table 9.2) shows the reverse, being lower than calculated, and substantially lower than new office buildings. For lighting, this is explained by the excellent daylight environment in the rooms, together with dimming controls and efficient sources and luminaires. The low consumption for fan use, together with very high air quality, underlines the effectiveness of the hybrid ventilation scheme and its spatial ventilation efficiency. Figure 9.9 shows the breakdown of fan energy use, and draws attention to the high proportion for toilets. Bearing in mind the intermittent use of toilets, this suggests that a more sensitive control might make further savings. It is reassuring to see that night ventilation (which successfully provided daytime comfort temperatures) represented a very small energy cost.

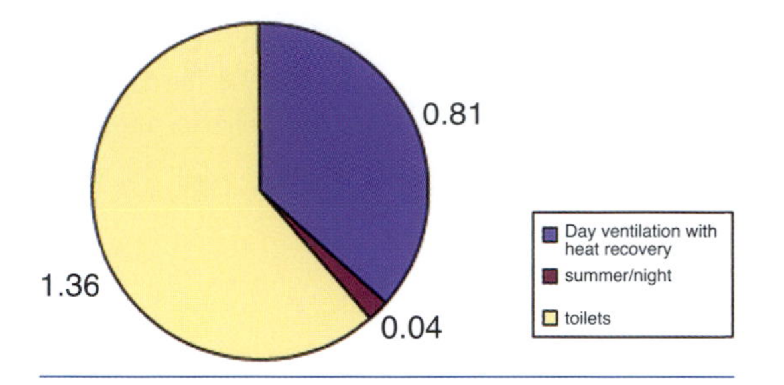

Figure 9.9 Electricity for daytime ventilation, summer night ventilation, and toilets

Note: Units are kWh/m²yr.

Comfort

Objective measurements show that the prevailing winter and spring temperature in the rooms was 20–22°C. Vertical gradients were no larger than about 2°C. The results of occupant survey questionnaires are shown in Figure 9.10.

These results show quite a high level of dissatisfaction. In particular winter heating comfort was poor with 25 per cent reporting high dissatisfaction. This was related to over-ventilation of the double skin void, already referred to, leading to low air temperature and draughts from the fresh-air inlets. It is not clear why 40 per cent reported dissatisfaction with air quality, since the objective measurements suggested that it was high.

The strong dissatisfaction with the acoustics was due to a very specific problem. Small holes in the second skin resulted in wind-driven resonance producing an irritating sound. This has now been corrected.

Finally, there seems to be clear agreement that the daylighting is very good, although since in the question daylight and view had been combined, it is not clear if the visual obstruction of the second skin and shading devices has caused some negative impact on view.

Conclusions

The Albatros project has demonstrated the viability of the application of a second skin to an over-glazed, leaky and poorly insulated 1970s tower block. The monitoring has, as is often the case, shown up a number of malfunctions which have had some serious effects both on energy use and user satisfaction. However, the causes have been identified and corrected and the building should now return a much better performance.

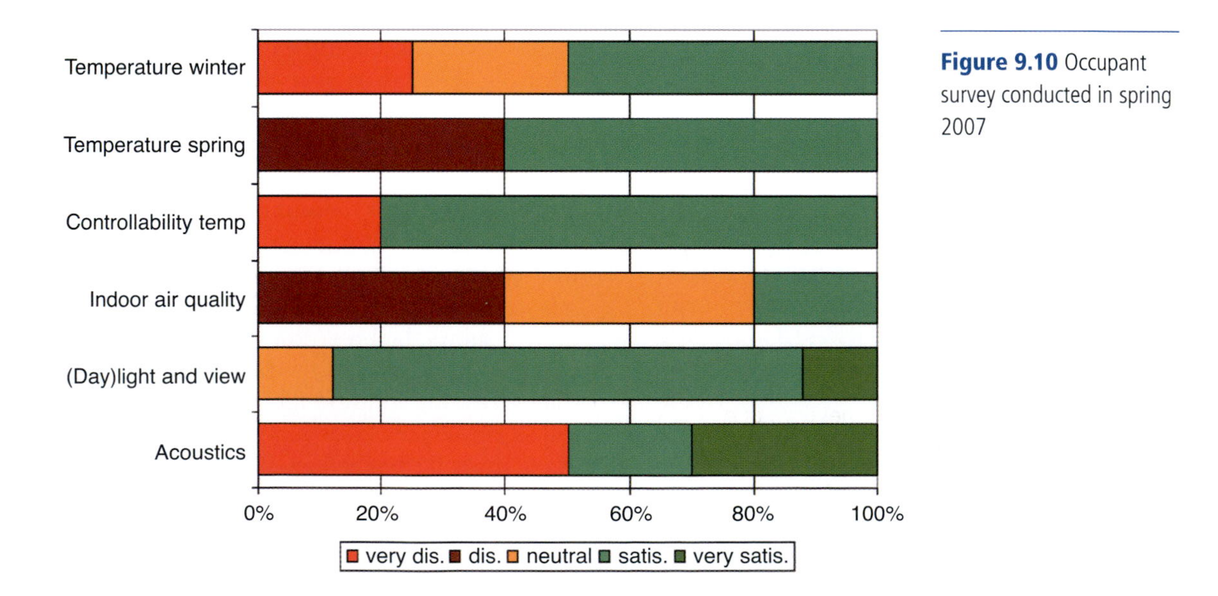

Figure 9.10 Occupant survey conducted in spring 2007

10 Lycée Chevrollier, Angers, France

Major fabric renovation improves insulation and shading to 33,000m² building

The Chevrollier High School was built in 1959. This large school houses 500 teachers, and 2700 students from sixth grade to school leaving age, providing both academic and vocational education. The building had never been refurbished and a number of problems were evident:

- Overheating in the south-facing classrooms.
- Degraded poorly fitting windows resulting in high infiltration.
- Uninsulated fabric and poor heating controls resulting in thermal discomfort.

Other problems relating to the fabric of the building included failure of some reinforced concrete elements, the degradation of finishes, and the presence of asbestos in certain areas. Furthermore, there was no mechanical ventilation system installed in the classrooms, now mandatory in French schools.

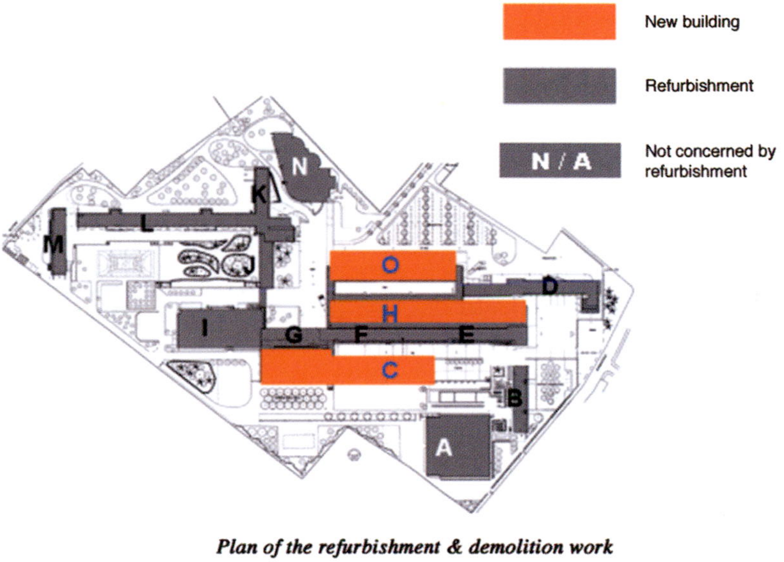

Figure 10.1 The original south elevation of the Lycée Chevrollier, Angers, France

New building

Refurbishment

N / A Not concerned by refurbishment

Plan of the refurbishment & demolition work

Figure 10.2 Plan of the refurbished and newbuild work

Strategy for sustainable refurbishment

The shallow plan of the building results in a very large surface area and this, together with high U-value and infiltration rate, made improvement to the thermal performance of the envelope a high priority. However, the shallow plan also means that there is good availability of daylight over the floor plan. This could realize electrical energy savings due to reduced artificial lighting loads, provided appropriate control systems were incorporated. Finally the problem of overheating in the south-facing classrooms needed to be addressed.

Main low energy measures

A. Thermal

- Externally applied insulation to east, north and west, internally applied to south wall reducing the wall U-value from 2.0W/m^2°C to 0.39, insulation to roof from 3.5 to 0.23, and floor, where crawl space allowed, 2.1 to 0.16.
- High performance low-e double glazing in thermal break aluminium frames with a U-value reduction from 6.0 to 1.9W/m^2°C.
- Reduced infiltration rate due to replaced windows and fabric cladding.
- New low NO_X emission gas boilers.
- New unheated atrium between original building and new block.

B. Lighting

- Improved daylight distribution in third floor classroom due to secondary lighting from north facade.
- Daylight and occupancy sensitive lighting controls.
- High efficiency sources and luminaires.

C. Comfort: Shading and ventilation

- Shading of whole south facade with fixed external louvres.
- Mechanical extract ventilation with passive inlets controlled by BEMS, with night ventilation.

D. Other features

- Photovoltaic array installed on roof of new block.
- Use of low volatile organic compound (VOC) emission finishes.
- Management (sorting) of building waste to permit recycling.

Insulation

The original envelope was comprehensively insulated as shown in Table 10.1. All cold bridges were insulated by external insulation except for the south facade where the visual intrusion on the exposed original facade was considered to be unacceptable. Floor insulation was possible for only a relatively small proportion of the building, where

Table 10.1 Specification of envelope insulation before and after refurbishment

	Before refurbishment		After refurbishment	
	composition	U-value W/m^2°K	composition	U-value W/m^2°K
wall	sandwich panel	> 2.0	W,E,N – external S – internal insulation	0.385
roof	no insulation	> 3.5	insulation	0.233
glazing and frames	single glazed steel frame	> 6.0	low-e double aluminium thermal break	1.9
floor 1 on soil	no insulation	2.0	no insulation	2.0
floor 2 over crawl space	no insulation	2.4	insulation underfloor	0.16

Note: All cold bridges treated except on south facade.

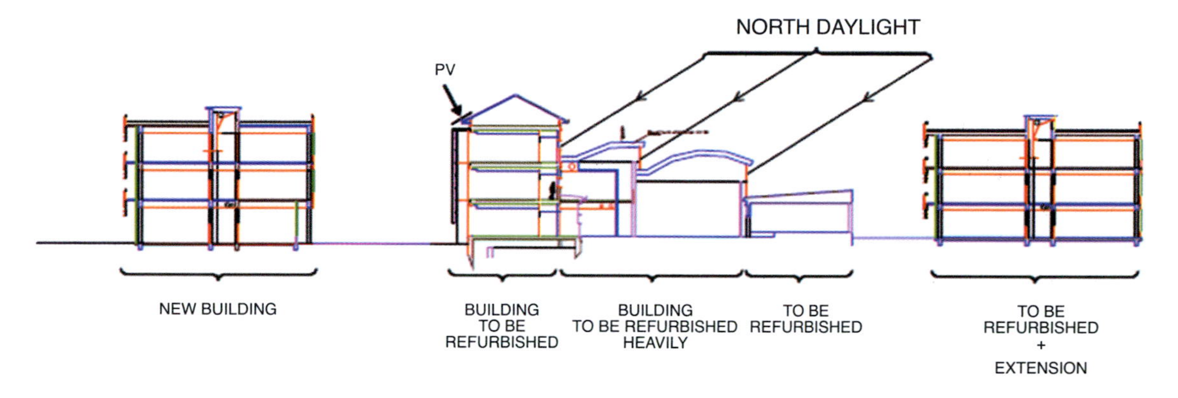

Figure 10.3 Diagrammatic section showing daylight and shading strategy

there was sufficient crawl space to give access to attach insulation to the underside of the floor.

Daylight and artificial lighting

The overall daylight strategy is illustrated in Figure 10.3. It shows the provision of extensive solar protection on the south facades of the new south block and the long refurbished block. It also shows the access to north light to the corridor of the refurbished block, and to the two new workshop blocks via north-facing roof lights.

Although the original classrooms (south side of old three-storey block) had adequate areas of glazing, the daylighting quality was poor. This was because there was a high illuminance gradient from the glazed facade to the back of the room (5.5 metres), exacerbated by the lack of shading other than internal curtains. The drawing of curtains to reduce glare interfered with ventilation and obstructed view, and often resulted in the lights being switched on.

Two major interventions were made. Firstly the installation of 8500m² of fixed horizontal louvre shading onto the outside of the south facade. This was first tested over a small area, and following the successful test was installed over the whole facade without significant modification. It has a major visual impact on the building. In addition to this, the replacement glazing was high performance selective transmittance type with a total transmittance of 40 per cent.

The shading geometry is such that no direct sunlight penetrates the facade from March to September. This provides an effective shading transmission coefficient of about 25 per cent. The aluminium louvres are finished on both upper and lower surfaces with an off-white paint of high reflectance. This causes some of the intercepted sunlight to be inter-reflected between the louvres and redirected to the back of the room, improving the daylight distribution.

The second intervention (on the top floor only) was to reinstate and enlarge secondary daylighting openings in the back wall adjacent to the corridor which itself was lit by high level clerestory windows. It is well established that a small increment of daylight factor (DF) where the DF is lowest can have a considerable benefit. This is because of the reduction in DF ratio.

Finally, all of the new room surfaces were of high reflectance. This leads to a high internally reflected component (IRC) of the daylight, also contributing to reducing the DF gradient across the room, as well as increasing the average illuminance level.

Artificial lighting

The light fittings in the refurbished classrooms are high efficiency fluorescent sources in low-glare luminaires complying with French Codes of Practice for educational buildings. The design illuminance is 500 lux.

In order to realize the energy saving potential of daylight, artificial lighting must be switched off when daylight is available. Manual switching in multi-occupancy rooms is very ineffective in this role, so light and occupancy sensitive controls have been installed.

The luminaires are arranged in two banks running parallel to the window wall. They are controlled by a manual switch – switching the whole room, with an occupancy detector, and an illuminance detector controlling each bank separately. The latter is to respond to the decreasing DF with distance away from the window.

The control logic is as follows:

room unoccupied	lights always off
room occupied/daylight	lights under on/off control in banks (initial manual switch-on)
room occupied	lights on (manual switch-on)

Performance

In most classrooms the perceived daylight quality is excellent. There are some exceptions in block F on the ground and first floor where the overshading by the new block C, together with the obstruction of high angle sky by the shading louvres, has lead to a low average value of DF.

Unfortunately the consumption of lighting has not been separately metered, and so it is not possible to disaggregate the contribution of the daylighting and the control system. There have been some problems with setting up the photosensitive control. The on/off control has caused annoyance to the occupants, to the extent that in some cases, teachers have drawn the curtains to ensure that the artificial lighting stays on. This problem will be eliminated when the intended dimming control is installed.

Ventilation

No special provision was made in the original classrooms for ventilation other than openable windows. In winter, usually windows were shut, but the high level of infiltration through the badly fitting steel windows gave sufficient fresh air to maintain a reasonable air quality. However, this infiltration was uncontrolled, and led to over-ventilation and subsequent waste of heat during times of high wind and when the room was unoccupied.

The upgrading of the envelope included new airtight windows, thus removing this 'fail-safe' background ventilation. This was replaced by a hybrid mechanical ventilation system under the control of the BEMS. In each classroom there is mechanical extract, with passive inlets under BEMS control.

In principle the system can respond to varying demands made on ventilation rate:

Status	ac/h
Unoccupied	0
Occupied	1
Heavily occupied	3
Night ventilation	6

The night ventilation is controlled as follows:

Night Vent	ON	between 2200 and 0600 hours
	IF	internal temp > 19°C
	AND	internal temp > external temp + 2

The cool night air can make contact with the thermal mass of all wall surfaces except the inside of the south facade (due to internal insulation), and the ceiling slab where there is a suspended ceiling. The floors are treated with sheet vinyl and linoleum also permitting good thermal coupling with the floor slab.

Performance

Figure 10.4 shows the evolution of temperature and CO_2 for one of the classrooms in block E, in January. The response of both, with the CO_2

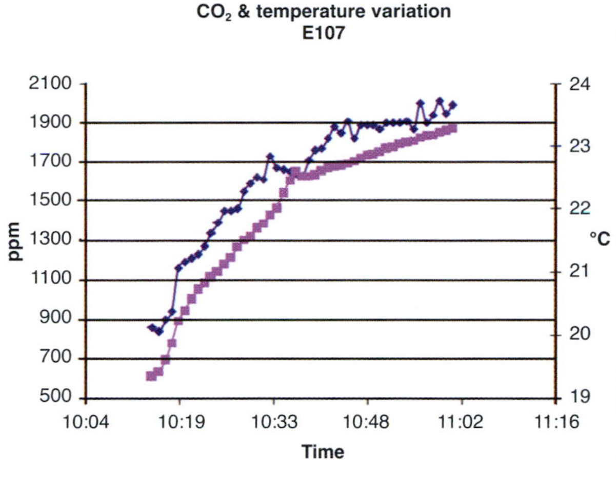

Figure 10.4 Temperature and CO_2 concentration during one hour of occupancy of classroom in building E

going far above acceptable levels, suggests that the ventilation is not operating correctly.

Generally, the air quality and temperature control is poor. This points to inadequate ventilation. The mechanical system is specified to deliver 3ac/h when the classroom is fully occupied. This gives a maximum fresh air per student of 3.5l/s compared with a recommended minimum of about 9l/s. Although there are openable windows, safety regulations prevent them from being opened widely.

The atrium

The atrium, Figure 10.5, is located between the old building G, and the new building C which forms the south and west walls of the atrium. The east wall and the roof are predominantly glazed. The east also contains the bridge linking the old and new blocks at high level.

At ground level, the atrium includes the undercroft beneath the new building C and the old building G, and thence to the outside via large glazed industrial type doors (Figure 10.6). In summer these are open, together with large vents in the roof under BEMS control. There is no shading, but these large openings are sufficient to keep summer temperatures to within acceptable limits.

Figure 10.5 The atrium formed between new building C on the left and refurbished block G on the right

In winter, the large doors and the roof vents will be kept closed. No mechanical heating is provided, but the temperature is elevated above the outdoor temperature due to heat losses from the surrounding buildings, and solar gains if available, and never falls below 10°C. This results in reduced heat losses from the adjacent buildings.

Artificial lighting is provided by eight spotlights directed up onto downward facing reflectors. There is some concern that this is a wasteful system due to a large proportion of the light missing the reflector.

The atrium has an important social function, acting as a meeting space, information point, and on occasions a space where events and performances take place. This is a good example how a low cost space (no servicing, little new structure) can be of great value.

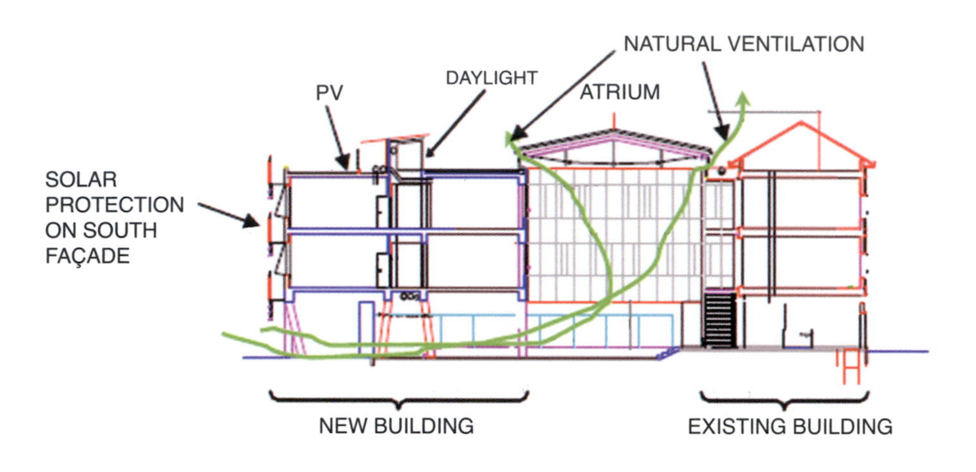

Figure 10.6 Ventilation openings in the roof of the atrium, under BMS control, and large openable doors at ground level give high rates for summer ventilation

Photovoltaic panels

The panels (78m^2) were originally to be mounted on the roof of the atrium to provide shading, but after technical problems it was decided to move them to the roof of block C (Figure 10.7). Their output was monitored from 17 July 2006 to 7 April 2008 and corrected for an average year.

Table 10.2 Performance of PV installation

Surface area m^2 PV panel	Total energy kWh/y	Energy/m^2 kWh/y	Total CO_2 displaced kg/y
78	7666	98	3449

Waste management and other environmental issues

During the demolition that was part of the refurbishment, a lot of waste material was produced, including hazardous waste such as asbestos. Wastes produced were recorded and segregated by type, allowing efficient recycling where possible, and efficient disposal where not.

The design and specification of all new components (both in the new building and the refurbishment) were chosen to minimize on-site waste, and with a view to their own recycling at the end of their life. The waste generated by the building in use was also considered and the design provided special recycling areas for valuable wastes such as paper and aluminium.

The local environmental impact of the materials used in the new building and the refurbishment were also given special consideration. Long-life finishes that required minimum chemical cleaning, and low VOC emission materials such as linoleum were specified.

Figure 10.7 78m^2 of photovoltaic panels installed on the roof of the new block C

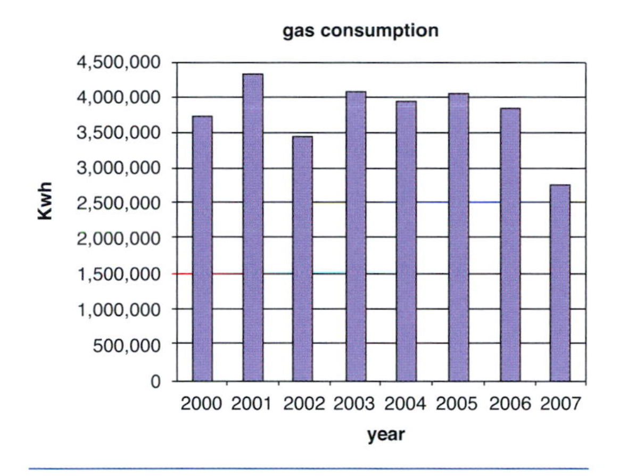

Figure 10.8 Total gas consumption 2000 to 2007. Includes estimated 240,000kWh for cooking

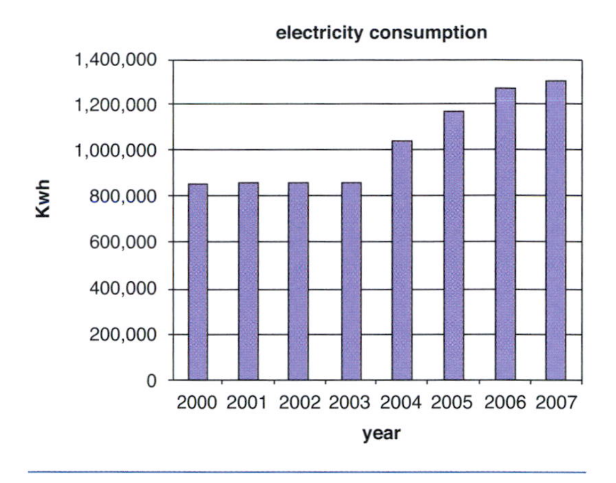

Figure 10.9 Total electricity consumption for 2000 to 2007

Overall energy performance

Gas consumption

The annual gas consumption in Figure 10.8 has been corrected for the variation in degree days from 2000 to 2007, and is quoted per m^2 to take account of the increase of heated space. It shows a significant reduction of 31 per cent for the year heating season 06/07.

Bearing in mind the dramatic reduction in fabric heat loss, this is rather less than may have been expected. There are two probable explanations. Firstly there have been problems with heating controls in some classrooms following damage and failure of the local radiator thermostats. This has led to overheating and then wasteful ventilation by window opening. Thermostats are being changed to make them more robust and tamperproof.

The other likely cause, common to many refurbishment projects, is that the heating standards have improved. This is not normally accounted for when comparing before and after, since, as in this case, temperatures and thermal satisfaction were not recorded in the original building.

Finally, the effect of the improved fabric insulation to blocks E and F, where the largest changes were made in fabric insulation, have been diluted by heating in other areas which have had less improvement, and other gas consuming activities such as cooking which have not changed. When this is taken into account, it is estimated that the reduction for blocks E and F alone would be about 45 per cent.

Electricity consumption

The annual electricity consumption in Figure 10.9 shows a relentless increase. There are several explanations for this. Firstly the area has increased by nearly 30 per cent. Secondly the small power loads have increased due to the installation of computers, multimedia equipment, machine tools and general electrical equipment. There must also be an energy penalty for the increased use of mechanical ventilation. These will mask the effect of the reduced electrical consumption for lighting, which unfortunately was not metered separately.

CO_2 emissions

The emissions have been calculated using a European average value for the CO_2 conversion factors for kWh to kg CO_2. These are:

1kWh electricity	0.44kg CO_2
1kWh gas	0.19kg CO_2

Table 10.3 shows the comparison for 2000 and 2007.

Comfort

450 questionnaires were circulated to the students. 57 per cent were returned and the results are shown in Table 10.4. These figures are for the whole building including the new part. It shows that satisfaction is 65 per cent and 55 per cent for winter thermal comfort and ventilation, but drops to 47 per cent and 44 per cent in summer. Satisfaction with daylight is high, at 85 per cent,

Table 10.3 CO_2 emissions for 2000 and 2007

2000	kWh	Emission	kg CO_2
electricity	860,332	0.44	378,546
gas	3,726,436	0.19	708,022
		total	**1,086,568**
2007			
electricity	1,305,603	0.44	574,465
gas	2,746,606	0.19	521,855
		total	**1,096,320**

Table 10.4 Post occupancy evaluation from the students for the whole school

Thermal comfort	Winter satisfaction	Yes	167	65%
		No	88	34%
		NA		1%
	Summer satisfaction	Yes	121	47%
		No	122	47%
		NA		6%
Ventilation	Winter satisfaction	Yes	142	55%
		No	109	42%
		NA		3 %
	Summer satisfaction	Yes	114	44%
		No	129	50%
		NA		6%
Natural lighting	Satisfaction	Yes	219	85%
		No	33	13%
		NA		12%
	Curtain operation	Yes	155	60%
		No	90	35%
		NA		5%
Artificial lighting	Satisfaction	Yes	119	46%
		No	134	52%
		NA		2%

Table 10.5 Post occupancy evaluation from teaching staff

			Build C		Build E		Build F		Build H		Global	
	Number		14		15		3		2		36	
Thermal comfort	winter satisfaction	YES	11	79%	10	67%	1	33%	0	0%	22	61%
		NO	3	21%	5	33%	2	67%	2	100%	14	39%
	summer satisfaction	YES	8	57%	5	33%	2	67%	0	0%	16	44%
		NO	5	36%	8	53%	1	33%	1	50%	16	44%
Ventilation	winter satisfaction	YES	7	50%	4	27%	1	33%	2	100%	12	33%
		NO	6	43%	10	67%	2	67%	0	0%	22	61%
	summer satisfaction	YES	5	36%	5	33%	0	0%	0	0%	7	19%
		NO	7	50%	8	53%	3	100%	1	50%	24	67%
Natural lighting	summer satisfaction	YES	11	79%	13	87%	3	100%	2	100%	30	83%
		NO	3	21%	2	13%	1	33%	0	0%	6	17%
	curtain operation	YES	6	43%	11	73%	1	33%	0	0%	20	56%
		NO	5	36%	3	20%	2	67%	1	50%	11	31%
Artificial lighting	satisfaction	YES	7	50%	9	60%	1	33%	0	0%	18	50%
		NO	7	50%	6	40%	2	67%	2	100%	18	50%

but drops to 46 per cent for artificial lighting, probably reflecting the problems with the photo-responsive controls.

Unfortunately, the student survey was not carried out building by building. However, this data is available from the teachers only, and is presented in Table 10.5. It follows a similar pattern but with much stronger criticism of the ventilation and summer comfort in blocks E and F. This is consistent with the CO_2 and temperature measurements shown in Figure 10.3.

Conclusions

The impression on visiting Lycée Chevrollier is that you are visiting a new building. The level of refurbishment is so high, that one is not aware of where the original building finishes and where the 30 per cent increase in area begins. Clearly there has been a huge investment in money and embodied CO_2, yet there has also been a great saving in all of the original CO_2-rich materials such as concrete and steel.

The CO_2 emissions for 2007, calculated using the standard European average, show no reduction from the 2000 value. However, the building is 30 per cent larger, and undoubtedly is serviced to a higher standard and contains far more energy-consuming equipment.

The difficulties with commissioning, which undoubtedly explain the poor performance in certain areas, are often experienced in new buildings and are not related specifically to refurbishment. Most of these problems have now been identified and will be corrected in the near future. As is usual with newly commissioned buildings, we can expect a steady and significant increase in performance as these problems are solved.

11 Daneshill House, Stevenage, UK

Phase change material installed in lightweight office interior to increase thermal mass

Daneshill House is a purpose-designed local government office building of 6898m^2 gross area, the larger part constructed in the 1950s as part of the Stevenage 'new town' development. The old block is a concrete-framed building originally clad with single glazed curtain walling, with openable windows. It was refurbished in the 1980s and 67 per cent of the cladding which was single glazed had its thermal insulation improved with secondary glazing and the outside pane replaced with tinted glazing in an attempt to reduce solar gains. The opaque part of the wall (33 per cent) was insulated to give a U-value of 0.45W/°Cm^2. However, this part still suffers from severe overheating in summer (Figure 11.1).

Figure 11.1 The Borough Council Offices built in the 1950s as part of the New Town of Stevenage. The original block (on the right) was refurbished in the 1980s and a new block added (left).

The new block, built in the 1980s, has a concrete frame and slab, and insulated brickwork cladding with a U-value of 0.45W/°Cm^2. The small windows have high quality double-glazed units and internal solar blinds.

Strategy for refurbishment

Three distinct areas of the building were considered – general offices in the old block, offices in the new block, and a Customer Service Area, also in the new block. The main objective was to address the problems of overheating in the old block. In common with many buildings of this type, summer overheating was a serious problem, exacerbated by ever increasing gains from equipment, and a series of above average summer temperatures. Air-conditioning had been installed in the new block from new, and there was considerable pressure to retrofit air-conditioning in the old block.

Another area of concern in the old block was the electric underfloor space heating system, which apart from poor control, had, due to its energy source, very high CO_2 emissions. It was replaced with a gasfired radiator system. Daylighting controls were also introduced in all office areas together with an innovative light emitting diode lighting scheme in the Customer Service Area. Finally the electric domestic hot water system was replaced by a gas boiler, together with a solar water heating array.

Figure 11.2 Stevenage Borough Council Offices interior after refurbishment

Main innovative energy saving features

The CoolDeck system

The open plan, carpeted floor and suspended ceiling in the old block did not lend itself to conventional passive cooling options of night cooling, since the thermal mass of the floor slabs was isolated from the interior. This called for an unconventional approach, where thermal coupling with the deck was achieved by fans drawing cool night air into the ceiling void, as in Figure11.3. In addition, the effective thermal mass was increased by including phase change material (PCM).[1] This system is referred to as CoolDeck (Figure 11.4).

Figure 11.4 The CoolDeck system: Air is drawn between the concrete slab and metal trays containing PCM

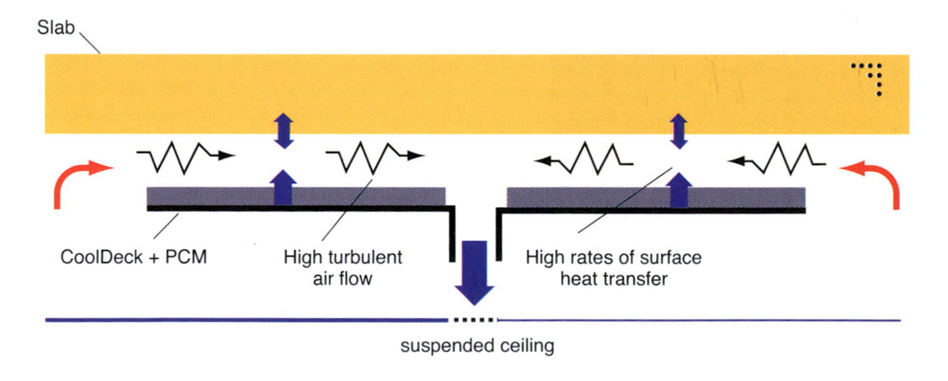

Figure 11.3 The thermal mass is coupled to the room air by fans drawing air over the concrete slab in the ceiling void. At night, outside air cools the room and slab. In the day, room air is circulated via the cooled slab

1 *Phase change* refers to the melting or solidifying of a material. Phase change materials behave as if they have very high thermal mass at and around their melting temperature.

Figure 11.4 shows air being drawn between the concrete slab and the metal trays containing PCM. The turbulent flow increases the surface heat transfer between the air and the slab and PCM. At night-time, outdoor air is delivered to the room by window mounted fans, and then drawn into the ceiling void to cool down the slab and 'freeze' the PCM. In the daytime the room air is cooled by flowing over the slab and PCM.

The trays cover only 15 per cent of the slab area but it is calculated that by including the PCM the available thermal mass has been effectively doubled.

The PCM is based on calcium sulphate and water, with additives to 'fine tune' the transition temperature to 24°C. It is held in plastic sachets which can be removed from the metal trays, as shown in Figure 11.5.

Figure 11.5 Plastic sachets containing PCM

Performance

The summers of 2003 and 2005 had temperatures well above average. The CoolDeck system alone did not lower temperatures sufficiently and complaints of discomfort were prevalent after sequences of hot days, when night-time temperatures did not drop below 22°C. Simulation with up to date climate data showed that it was impossible to meet the Chartered Institute of Building Services Engineers (CIBSE) criterion, of less than 1 per cent occupied hours above 28°C, with CoolDeck alone.

It was decided to add top-up cooling to the air by local direct expansion (DX) chilling units located in the ceiling void (Figure 11.6) to

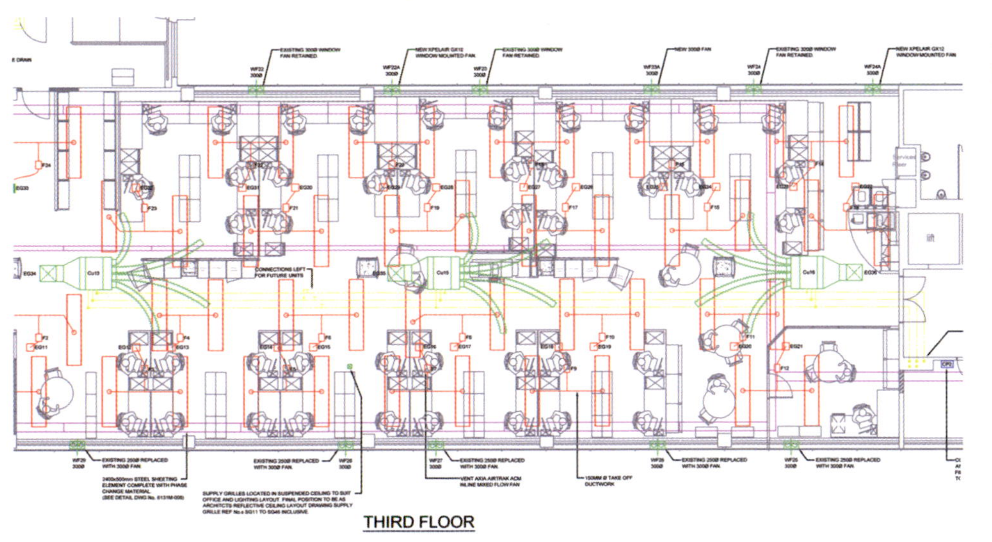

Figure 11.6 Floor plan showing the DX units (green) and distribution ducts providing cooling top-up

Table 11.1 Comparison between conventional air-conditioning, CoolDeck, and CoolDeck plus top-up cooling

Air-conditioning (1)		CoolDeck (2)				CoolDeck with top-up (3)			
annual		annual		difference		annual		difference	
kWh	CO_2(kg)	kWh	CO_2(kg)	CO_2(kg)	%	kWh	CO_2(kg)	CO_2(kg)	%
17,512	7530	3468	1491	6039	80	4468	1921	5609	74

ensure that the temperature of the PCM and the slab were lowered to below the transition temperature. The installed capacity was about 25 per cent of a conventional air-con system.

This 'peak lopping' was very effective, showing a large drop in complaints of overheating. Table 11.1 shows a comparison between conventional air-con electricity consumption, CoolDeck only consumption, and the CoolDeck with top-up, showing a 74 per cent saving. These are calculated values, since due to commissioning problems, monitored data is not available.

Important aspects of the system are that the presence of the mechanical cooling is not evident and that its control is automatic. Occupants remain under the impression that the building is cooled by passive night ventilation, and are more tolerant of daytime temperatures. If mechanical cooling were more in evidence it is likely that demand for its use would become common.

Energy efficient air-conditioning controls

The offices in the 1980s block were already air conditioned. A design study of the ground, first and atrium floors, carried out in 2004, indicated that 42 per cent of the total energy used was fan energy, and predicted that a saving of 22 per cent could be made by the installation of automatic damper controls.

The new control strategy included three innovative features:

1. pre-conditioning periods;
2. an adaptive cooling set-point;
3. fresh air demand based on CO_2 concentration.

The pre-conditioning control logic is illustrated in Table 11.2.

The adaptive control takes account of the effective acclimatization that people show after periods of hot weather. Figure 11.7 shows the set-point responding to the outside air temperature.

The CO_2 concentration demand control strategy is shown in Figure 11.8.

Performance

Results for two short monitoring periods, before and after, have been extrapolated to indicate the annual performance, and are shown in Table 11.3. The new control strategy shows an average reduction in annual CO_2 emissions of 35 per cent. This demonstrates how low-cost changes in control strategy (and almost zero embodied CO_2) can make significant savings.

Energy-efficient lighting controls

High efficacy light sources and luminaires were installed in the second, third and sixth floors of the old building, together with photo-sensitive and occupancy detecting controls. Design illuminance is 400 lux with an installed capacity of 9.7W/m^2. The artificial lighting is on on/off control with a control point of 400 lux.

Luminaires are arranged in three banks each side, parallel to the glazed facades of the office, allowing successive switch-on from the central to the outside bank as the daylight diminishes. The two outer banks (each side) are controlled with the photo-sensor, and the two central banks are controlled by passive infra-red (PIR) occupancy detectors with a delay of 15 minutes.

Table 11.2 Control strategy for the new block air-conditioning

AHU 17 and 18 control options	Night (unoccupied) 18:00 to 06:00	Pre-conditioning 06:00 to 08:00		Day (occupied) 08:00 to 18:00	
	Normal	Morning pre-cool: summer	Morning pre-heat: winter	Normal operation	Economy mode (manual enable/disable)
AHU fans	Off	On	On	On	Off
Extract fans	Off	On	On	On	On
Air heating	Off (night setback controlled by perimeter heating)	Off	Heating set-point of 20°C	Heating set-point of 20°C	Off
Air cooling	Off	Off	Off	Comfort cooling set-point of 24°C adjusted based on adaptive profile	Off
Damper control	Closed	Full fresh air	Full recirculation	Modify to satisfy fresh air and free-cooling requirements	Full fresh air
Criteria	–	If at 06:00: space air temperature greater than 21°C AND greater than ambient temperature (plus 1°C fan pick-up). Ambient air temperature greater than 14°C	If at 06:00: space air temperature less than 21°C	Minimum supply air temperature of 17°C	Space air temperature greater than 21°C AND AHU not in cooling AND ambient air temperature is above 14°C AND CO_2 concentration below 700ppm

Note: AHU = air handling unit.

Performance

There were no data available for pre-installation performance; therefore measured performance is compared with Typical and Good Practice. This is shown in Table 11.4. The performance is disappointing, showing a 38 per cent improvement on a Typical office, but a 44 per cent increase over a Good Practice office.

There are several factors that could explain this disappointing performance. Firstly the installed illuminance of 400 lux is high, particularly for office tasks largely based on the use of VDUs. Secondly, on/off control is very wasteful, since it leads to no contribution being made by daylight until 400 lux is exceeded. Thirdly, there is much evidence to show that the control point can be lower than the design artificial illuminance, in this case, say 300 lux, since the spectral qualities of daylight allow for a lower working illuminance. Finally, the high design luminance

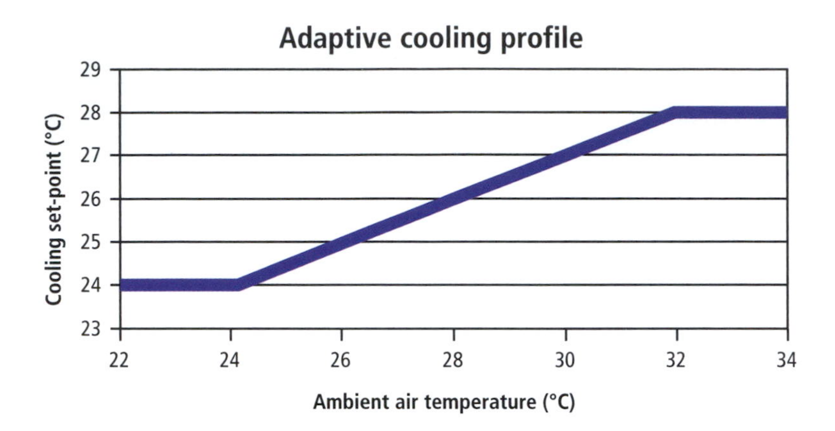

Figure 11.7 Adaptive cooling set-point used to control air-conditioning in old block

applies to the central zone (mainly circulation), and the 15-minute detection delay may well mean that this high level of illumination is on most of the time.

Photo controls are often found to be set up wrongly and it is essential to check with an illuminance meter that the correct switching thresholds are being achieved by the control system.

Light emitting diode (LED) lighting in Customer Service Centre

The Customer Service Centre is typical of the new breed of 'one stop shop' found in City Council offices, where residents come for a wide range of Council business. Some business may be stressful, and there may be protracted waiting periods.

Note 2: Percentage fresh air for damper modulation based on CO_2 concentration and free cooling
Dampers will modulate based on maximum fresh air requirement from the two schedules

Fresh air requirement based on CO_2:

CO_2 (ppm)	% fresh air
400	20
600	20
800	100
1000	100

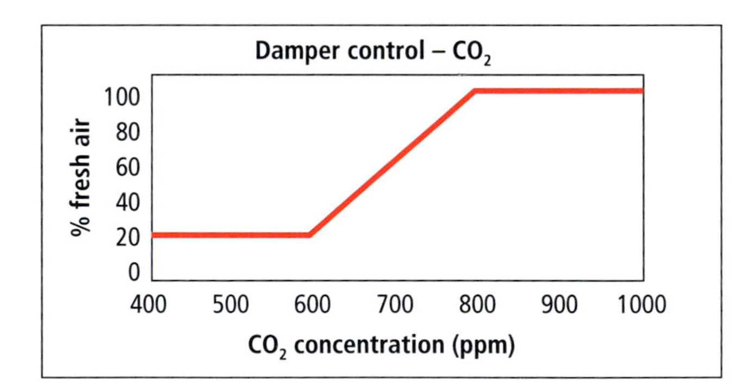

Fresh air requirement based on free cooling
Free cooling when Space air temperature > 21°C AND Ambient air temperature < Space air temp.
When ambient < 17°C: Percentage fresh air = 100 > Space air temp. –17 / (Space air temperature – ambient air temp.)
When ambient > 17°C: 100% fresh air

Figure 11.8 CO_2 concentration ventilation demand control strategy

Table 11.3 Annual CO_2 savings due to improved control strategy with CO_2 controlled demand and adaptive set-point

Ground floor atrium					**First floor atrium**				
Before		**After**			**Before**		**After**		
Monitoring dates					Monitoring dates				
20/06/05 to 06/07/05		23/05/06 to 15/06/06			07/07/05 to 21/07/05		09/04/08		
Annual		Annual		Diff	Annual		Annual		Diff
kWh	CO_2 (kg)	kWh	CO_2 (kg)	%	kWh	CO_2 (kg)	kWh	CO_2 (kg)	%
50,832	21,858	35,083	15,086	31	53,293	22,916	31,006	13,333	42

Note: Results are extrapolated.

The unusual lighting installation (Figure 11.9), which changes colour under the control of a timer, is to improve the visual quality and interest of the waiting areas. It also provides light at high luminous efficiency.

The energy consumption of the system was found to be 45 per cent above the Good Practice figure, which was disappointing. Monitored data showed that the system had been operated for at least three hours per day longer than necessary. This operational issue has now been corrected and the adjusted figure calculated now shows only a 9 per cent increase over Good Practice. It seems that although LED lighting offers potential high efficiency when used as the main white light source, decorative colour installations such as this tend to consume extra power due to the lower spectral sensitivity of the eye to the wavelength used.

Figure 11.9 The LED lighting system in the Customer Service Area is of high luminous efficacy and is programmed to make changes in colour to add visual interest to the waiting area

Table 11.4 Electricity consumption for lighting, third floor, 2007, compared with Typical and Good Practice

Year	**Area**	**Consumption**	**Occupation time**	**Annual consumption**	**Actual**	**Typical**	**Good practice**
	m^2	kwh/d	days	kWh	kWh/m^2 year		
2007	398	61	252	15,372	39	54	27

Table 11.5 Energy consumption of Customer Service Centre, actual, adjusted and Good Practice

Year	Area	Consumption	Occupation time	Annual consumption	Actual	Typical	Good practice
	m^2	kW	h	kWh	kWh/m^2year		
2007	198	2.1	3698	7766	39	54	27
Adjusted	198	2.1	2800	5880	**29.7**	**54**	**27**

Solar water heating array

Domestic water heating for the toilets and kitchens had been provided by two 800 litre 9kW calorifiers. Water usage was approximately 1000 litres per day. Energy consumption was also monitored and found to be about 85kWh per day, or 0.25kWh per person. In terms of CO_2 this is about 0.1kg per day per person, or a total of 11.2 tonnes per year.

The electric calorifiers were replaced with an array of high efficiency evacuated tube collectors with a condensing gas boiler backup. Two 500 litre storage vessels have twin heating coils with the solar circuit connected to the lower, and the gas boiler to the upper coil. This encourages stratification leading to higher collector efficiency.

The solar collectors, Figure 11.10, have an effective collector area of 18m^2 and the solar circuit is filled with 120 litres of fluid containing anti-freeze and corrosion inhibitor. The collectors are in three banks each with a flow rate of 4.5l/min. A Resol type ES controller compares collector temperature with the temperature in the calorifiers; the controller then switches on the appropriate variable speed circulating pump. An electronic display board, Figure 11.11, shows the temperature of the collectors, the stored hot water temperature, and the accumulated thermal energy.

Figure 11.10 The evacuated tube array of solar thermal collectors contributing to domestic hot water

Performance

Figure 11.12 shows the actual output for the solar hot water system together with the gas consumption. Hot water consumption is also shown. The high water consumption in May and November were due to a water leak and servicing work.

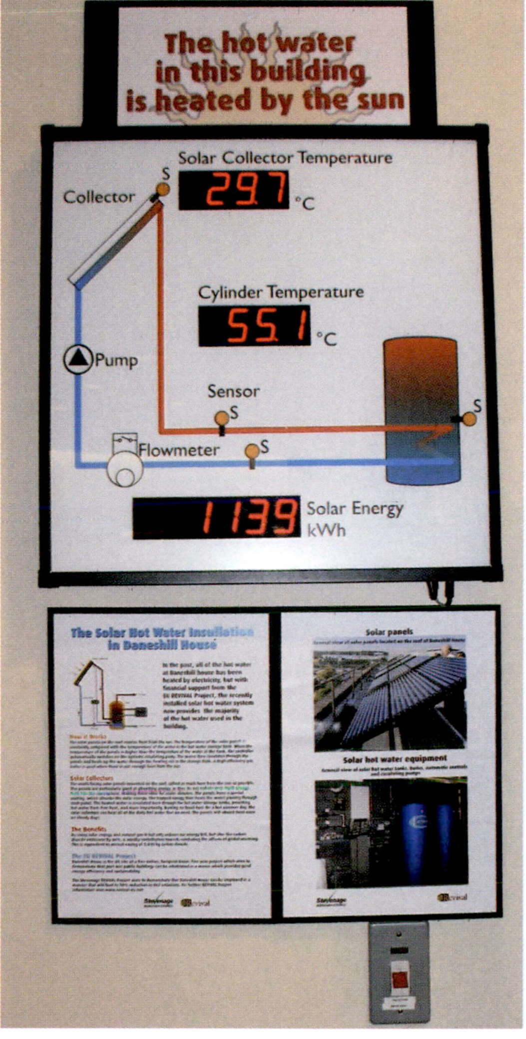

Figure 11.11 The electronic display board giving a real time readout of temperature and collected energy

Assuming the existing gas boiler is 90 per cent efficient, the solar energy collected reduced the gas demand by 28 per cent.

Increased space use efficiency

The energy used by a building is roughly proportional to the area that is conditioned. Hence performance and benchmarks are quoted in kWh/m^2. However it is the number of people that is the key parameter. An unoccupied workstation, or an empty meeting room, or wasteful over-provision of space per person, is a waste not only of capital resources but also of energy.

By changing furniture design and layout, and moving to paperless archiving, the mean occupancy density in Daneshill House has been increased; that is, the space provision reduced from 9.2m^2 per person to 7.6m^2 per person. This alone would reduce the CO_2 emissions per person by 17 per cent.

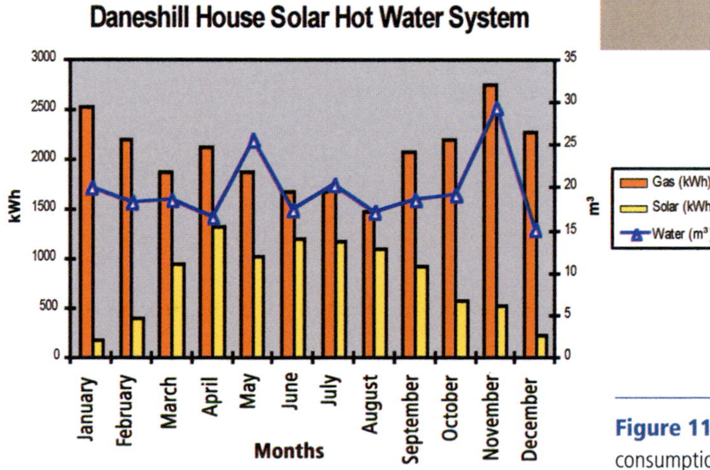

Figure 11.12 Collected solar energy, gas and water consumption 2007

Table 11.6 CO_2 reduction by solar system from original electrical heating, and compared with all-gas system

	Unit	All electric	Gas+ solar	All gas	Gas + solar
Solar input	kWh		9707		9707
Gas input	kWh		24,889	34,595	24,889
Elec. input	kWh	25,926			
CO_2 emissions	kg	11,148	4729	6573	4729
Annual CO_2 saved	kg		6419		1844
Annual CO_2 reduction	%	Compared with elec >	58	Compared with all gas >	28

Post occupancy evaluation

Figure 11.13 shows the results of the occupant survey from 279 respondents. The most interesting conclusion is that the refurbished floors with CoolDeck returned the highest level of satisfaction, higher than refurbished air-conditioned rooms, the worst of which recorded a dissatisfaction vote of 69 per cent.

It is also important to note that the highest satisfaction was recorded on the third floor of the old block, which was where the higher density layouts had been adopted.

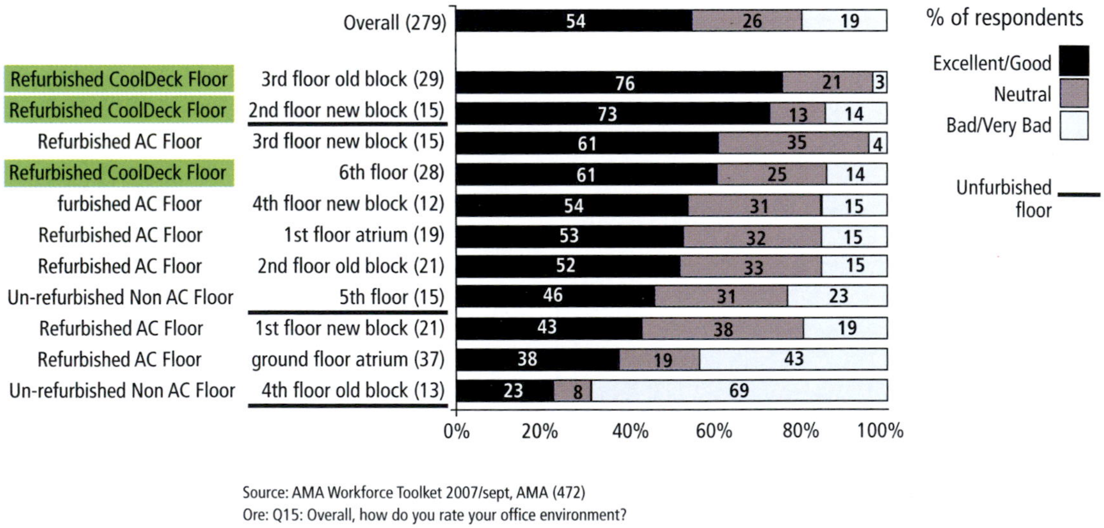

		Excellent/Good	Neutral	Bad/Very Bad
	Overall (279)	54	26	19
Refurbished CoolDeck Floor	3rd floor old block (29)	76	21	3
Refurbished CoolDeck Floor	2nd floor new block (15)	73	13	14
Refurbished AC Floor	3rd floor new block (15)	61	35	4
Refurbished CoolDeck Floor	6th floor (28)	61	25	14
furbished AC Floor	4th floor new block (12)	54	31	15
Refurbished AC Floor	1st floor atrium (19)	53	32	15
Refurbished AC Floor	2nd floor old block (21)	52	33	15
Un-refurbished Non AC Floor	5th floor (15)	46	31	23
Refurbished AC Floor	1st floor new block (21)	43	38	19
Refurbished AC Floor	ground floor atrium (37)	38	19	43
Un-refurbished Non AC Floor	4th floor old block (13)	23	8	69

0% 20% 40% 60% 80% 100%

% of respondents

Excellent/Good
Neutral
Bad/Very Bad
Unfurbished floor

Source: AMA Workforce Toolket 2007/sept, AMA (472)
Ore: Q15: Overall, how do you rate your office environment?

Figure 11.13 Results of the occupant survey from 279 respondents. The most interesting conclusion is that the refurbished floors with CoolDeck returned the highest level of satisfaction, higher than refurbished air-conditioned floors, the worst of which recorded a dissatisfaction vote of 69 per cent.

Table 11.7 Energy and CO_2 emissions for gas and electricity, before and after refurbishment

	Gas	CO_2 kg	Electricity	CO_2 kg	Total CO_2
Before refurb	397,214	75,470	1137,058	500,305	575,780
After refurb	406,015	77,142	1353,451	595,518	672,660

Overall energy performance

Both heat and electricity demand have increased, by 22 per cent and 19 per cent respectively, and this is disappointing. However it must be borne in mind that the building had a serious overheating problem, and that the conventional solution to this would have been air-conditioning, which inevitably would have led to a much larger increase in electricity demand.

The increase in gas consumption is less surprising, since the underfloor heating and the domestic hot water, previously heated by electricity, are now heated by gas. Obviously the savings in electricity from these heating roles have been more than compensated by the extra electricity for the CoolDeck and other mechanical ventilation.

As with all newly commissioned buildings it is very likely that improvements in performance will continue to be made.

12 Ministry of Finance Offices, Athens

Retrofit shading and mechanical system improvements to large office building improve environment and energy performance

This large five-storey building (30,000m^2) was originally built in the 1960s as an engineering factory. In 1980 it was remodelled to form the central office of the Ministry of Finance, handling the taxation for all of Greece. At this time issues of energy efficiency and indoor environmental quality were not a priority. Recently, increasing dissatisfaction with the environmental conditions, together with a growing awareness of the need for good energy performance, have prompted this current work.

Refurbishment strategy

The main characteristic of this refurbishment is that it concentrates on systems and controls rather than fabric. This means that the costs, both in money terms and in embodied energy, are relatively low in relation to the expected savings. The only major investment in the fabric are the external shading devices and the replacement of the glazing on the north facade.

Figure 12.1 Main facade facing SSE

Figure 12.2 Newly fitted vertical and horizontal shading devices on WSW elevation on floors 2 and 3 with original concrete brise soleil on upper floors. Material is grey painted perforated steel to give the best combination of diffuse daylight and shading. Original internal louvre blinds shown on the right

Main energy saving features

Fabric improvements

- Installation of advanced shading devices to control solar radiation gains both in winter and summer.
- Replacement of single glazing with low-e double glazing on the north facade reducing heat loss in winter and heat gain in summer.

Night-ventilation techniques

- BEMS controlled night ventilation up to 8ac/h depending on indoor and outdoor air temperatures .

Ceiling fans

- Installation of specially designed ceiling fans which allow the extension of the comfort zone up to an indoor air temperature set-point of 28.5°C.

Daylighting and artificial lighting

- Use of photo-sensitive controls to maximize the efficient use of natural light.
- Utilization of high efficiency lamps and improved luminaires.

Heating

- Installation of circulating pumps of the inverter type controlled by the BEMS according to the internal and external temperature, as well as the temperature difference between inlet and outlet of the heating water through the energy meters, which are installed on each boiler. In every office a controller is installed giving the user an adjustment capability of +/–5°C over the operation of the fan coil unit.

Cooling

- Installation of a demand control cooling system per user. This system consists of a motorized valve which is controlled by a thermostat/transmitter combined with a valve controller. The controller is set on a minimum cooling temperature of 26°C. For the cooling consumption measurement an energy metering device is installed in the area of each user. In this way they will have the ability to control and monitor their own consumption. The reading of the energy meter will be transmitted to the central controller.

Ventilation

- Demand-controlled ventilation using CO_2 detectors to optimize and control indoor air quality.
- Selective use of fresh air whenever this is cooler than the recirculated indoor air, as well as redistribution of cooler air from nearby indoor spaces with different indoor comfort requirements.

- Heat/coolth recovery on ventilation, by a heat exchanger unit in the exhaust air stream. This provides pre-heating or pre-cooling to the incoming ventilation air, contributing considerably to the heating and cooling savings.

Energy management, control and monitoring

- Using radio controls, no cable installation is necessary which means that disturbance to the users is avoided. All the sensor outputs are transferred to the BEMS terminal via M-Bus protocols. The BEMS takes into account the consumed energy and the water volume, and controls the inverter pumps and also the activation sequence of the chillers.
- The BEMS gathers all consumptions and records and allocates the energy costs. With the system described above, there are two types of controls. One in each zone, which gives freedom to the user, within limits, and one in the heat/coolth production system which works according to the weather conditions, and according to the end users' demands.

Performance

Thermal comfort and air quality

These had been the key issues and sources of complaint prior to the refurbishment. Several measures adopted were aimed at improving comfort, without incurring further energy consumption. These measures include external shading, night ventilation, ceiling fans and the demand-controlled fresh air ventilation with CO_2 detection.

The success of these measures is indicated by objective measurements of temperature and CO_2 concentration, and subjective measurements by user questionnaires.

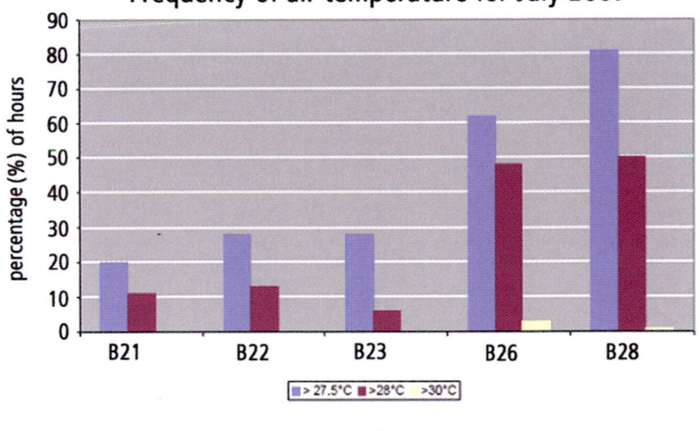

Figure 12.3 The frequency of temperatures in offices

Figure 12.3 shows the frequency of temperatures in five of the office areas for the hottest month, August 2007. The performance in offices B21, B22 and B23 was good, with no temperatures above 30°C recorded and only 10 per cent, 12 per cent and 6 per cent of occupied hours above 28°C. Areas B26 and B28 showed considerably higher temperatures with 3 per cent and 1 per cent above 30°C respectively. However, it must be borne in mind that the ceiling fans allow the comfort zone to be extended to 28.5°C.

Table 12.1 shows the ambient temperatures recorded on the site for June–August 2007, indicating an average of 32°C with a maximum of 37°C. Internal temperatures were kept consistently about 10°C below this.

As part of the energy efficiency measures, demand-control ventilation was installed to prevent wasteful over-ventilation, and prevent a lowering of air quality by under-provision of fresh air. Fresh air is also available by openable windows, although at times of high outside temperature this will lead to local discomfort and increased cooling load.

Figure 12.4 shows a graph of CO_2 concentrations for August in the monitored areas and it can be seen that concentrations above 1000ppm occur quite frequently. Specifically, CO_2

Table 12.1 Ambient temperatures for the site June–August 2007

	June			July			August		
	Max	Avg	Min	Max	Avg	Min	Max	Avg	Min
Daily max	43	30	22	41	32	28	37	32	26
Daily mean	32	24	18	33	27	22	31	26	22
Daily min	25	18	12	26	22	15	27	21	16

concentrations exceeded 1000ppm for 57 per cent of the time in June and July, 38 per cent in August. In the winter months the situation is slightly worse with it being exceeded for 58 per cent and 65 per cent in December and January respectively.

These high CO_2 concentrations were investigated in the autumn of 2007 and the ventilation rates were found to be low, as shown in Table 12.2. As a result of this investigation, fresh air ventilation rates were increased in February 2008 as indicated.

Table 12.2 Measured ventilation rates in the data entry rooms before improvements

	m^3/h	$V(m^3)$	l/s/pers
B 21	220	548	2.50
B 22	200	653	2.50
B 23	425	584	4.72
B 26	275	564	3.10
B 28	306	550	3.40

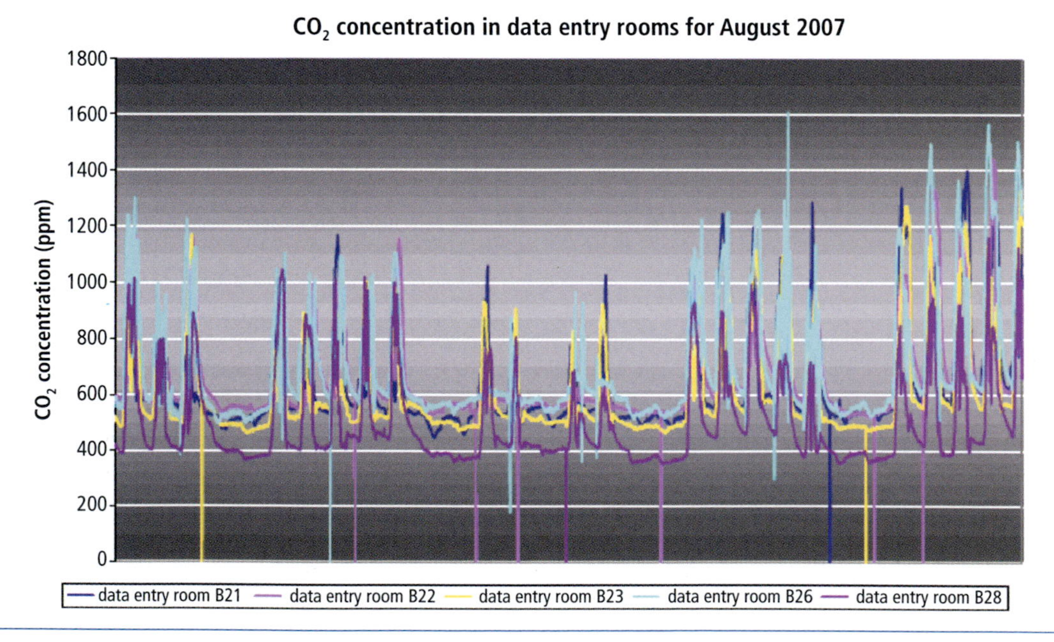

Figure 12.4 Monitored CO_2 concentrations for August 2007

Daylighting and artificial lighting

The original glazing ran in a continuous strip between cill level at 1.2m to the ceiling at 2.4m. Modelling studies using the software Radiance and Daylight had shown that average daylight factors (DF) exceeded 2 per cent in almost all cases, including rooms shaded by the large concrete brise soleil on the fourth and fifth floors.

The addition of the shading devices on the SSE, SSW and WNW, in order to control solar gains, inevitably reduces the available daylight. However, due to the geometry of the devices, they tend to shade direct sun, more than light from the diffuse sky. They also can reflect diffuse light back into the room. In this way, they reduce the non-uniformity of illumination, without reducing the average daylight illuminance below a level requiring artificial light.

A systematic study was carried out using radiance to find the best combination of reflectances, ranging from 80 per cent (nominally white) to 20 per cent. A value of 50 per cent reflectance (light grey) was chosen as the best compromise between illuminance, uniformity and reduction of glare.

The artificial light is provided by high efficiency Osram Lumilux T5 sources with high frequency dimmable ballast giving 104 lumens/W (compared with 75 lumens/W for conventional fluorescents). The lighting is controlled through the BEMS that receives signals from internal sensors in the rooms. Sensors were placed in two rows at 2.4m and 6m from the window wall (the room was 9m deep) and each sensor controlled the four surrounding luminaires.

Daylighting performance

The performance of a daylighting design must be judged by both its ability to displace artificial lighting energy, and to provide an acceptable luminous environment for the occupant. In terms of the former the energy consumption for lighting had been monitored before the refurbishment took place and found to be 46kWh/m^{2}yr. Consumption dropped to 38.6kWh/m^{2}yr, a reduction of 16 per cent.

This is a disappointing reduction and could be accounted for by the increased efficiency of the lamps alone. It could be due to over-illumination, although quoted illuminance measurements suggest that these are between 200 and 300 lux. However, these values were measured using the downward facing control sensors and it is not clear if the values were corrected for illuminance on the workplane. This could also explain the disappointing performance of the control system.

The photovoltaic array

The photovoltaic array was originally going to be part of the shading system. Technical considerations changed this to a roof-mounted array. The panels, total area 25.8m^2, are mounted at 30° inclination facing south, on the roof terrace of the building. Their operation has been recorded since the beginning of January 2008 for two months. The array had generated 447kWh of electricity.

Only 7 per cent and 10 per cent of the annual solar radiation is available in January and February. Assuming a constant efficiency, a predicted annual yield would be 2629kWh.

Overall energy performance

Table 12.3 shows that for the year 2000, the primary[1] energy consumption had been 53.6kWh/m^2 for heating and 87.9kWh/m^2 for cooling. For the year 2007, after renovation, the measured energy for heating was 34.3kWh/m^2, and for cooling, 47.0kWh/m^2.

The winter of 2007 was mild, and for com-

1 Primary energy is the energy value of the source fuel. For electricity the delivered energy is multiplied by 3.3.

parison the heating figure must be corrected to the same number of degree hours. This gives 38.2kWh/m^2 for the corrected heating energy, showing a reduction of 29 per cent. The cooling energy is reduced by 46 per cent.

Table 12.3 Primary energy consumption for heating, lighting and electricity

	2000	2007–08	% reduction
Annual cooling energy kWh/m^2	87.9	47.0	46
Annual heating energy kWh/m^2	53.6	38.2	29
Annual lighting energy kWh/m^2	46.0	38.6	16
Annual electricity kWh/m^2 (office equipment)	26	26	

Heating

Two observations can be made about the heating energy. Firstly, the actual value is still high – 38kWh/m^2 for a climate with only 4000 degree hours for base temperature 18°C (i.e. 167 degree days), and for a building with high internal gains. Secondly the reduction of 29 per cent is modest, bearing in mind the investment in low-e glazing, and demand-controlled ventilation.

The annual heating energy quoted above includes electrical power for fans and pumps. The probable explanation for the result is that since the actual heating load is small, the total energy is dominated by the electrical part, which has not changed very much. It is also an indicator of just how much parasitic energy can be used in air conditioning systems.

Cooling

The cooling energy has shown a significant decrease of 46 per cent. This must be due in part to the effect of the shading devices reducing solar gains. Further contributions must be made by the improved demand-controlled ventilation with variable speed fans. Reduction in lighting use due to the photo-sensitive controls should also reduce internal gains, but the relatively small reduction in lighting energy, suggests that that component is a relatively small one.

Comfort surveys

The subjective outcome of the conditions shown under *Thermal comfort and air quality* is indicated in Figure 12.5 in the bottom two rows. This includes the responses of all office areas. Individual areas showed slight difference, with the offices on the NNW orientation showing 60 per cent satisfaction with temperature, but the ENE orientation achieving only 33 per cent. It is interesting to note that the areas returning poor performance are those which are not usually associated with problems of solar gain. This suggests that the problem lies with the combination of high internal gains and poor control or inadequate capacity. This issue is already receiving attention.

Satisfaction with ventilation shows a similar pattern with only 40 per cent being satisfied overall and dropping to 23 per cent in the ENE office.

In interpreting these results it must be borne in mind that high levels of satisfaction are rarely reported. Typically people report neutrality when there is no adverse condition prevailing. There is also a probable link between the undisputed high CO_2 concentrations and the perception of thermal discomfort – that is, if CO_2 concentration and humidity are kept low, the higher temperatures may be more acceptable. Furthermore, the comfort survey was conducted in October 2007 before the increased ventilation rates had been implemented and before the ceiling fans had been installed.

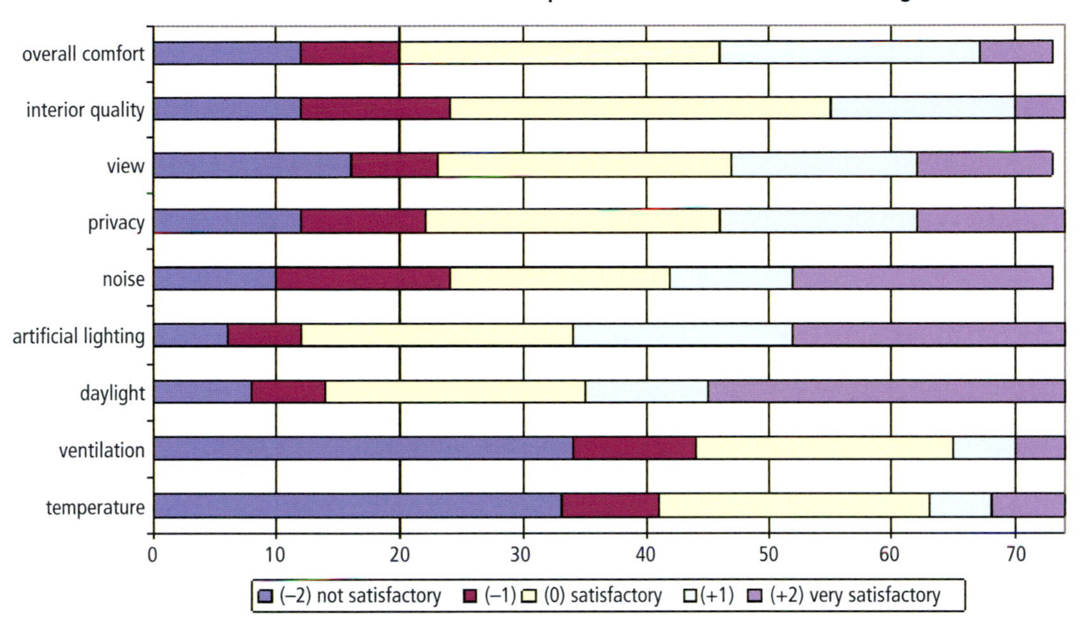

Figure 12.5 Comfort survey

13 The Meyer Hospital, Florence

Refurbishment and addition of large greenhouse to historic building in warm climate

The building dates from the 1930s when it was a hospital for tuberculosis patients. It consists of a long three-storey terrace running east–west, of traditional construction with very thick walls and a tiled roof. A central corridor serves rooms on the north and south side (Figures 13.1 and 13.2). The total area is 10,480m². The building had been semi-derelict for some years. A new paediatric hospital has been built on the site adjacent and the refurbished building forms the administrative and reception centre.

Refurbishment strategy

Being of traditional construction the building was non-insulated and single glazed. The first consideration was how to insulate the envelope. The roof presented no problem, since during reconstruction insulation material was included, in this case mineral fibre quilt. The windows were also replaced giving the opportunity to incorporate low-e double glazing in timber frames, giving a U-value of 2.85W/°Cm². This also reduced uncontrolled infiltration.

Figure 13.1 South elevation of the Villa Ognissanti, the original Meyer Hospital

The solid walls presented some problems. External insulation would have disturbed traditional detailing of this historic building, whilst internal insulation would have reduced thermal inertia.

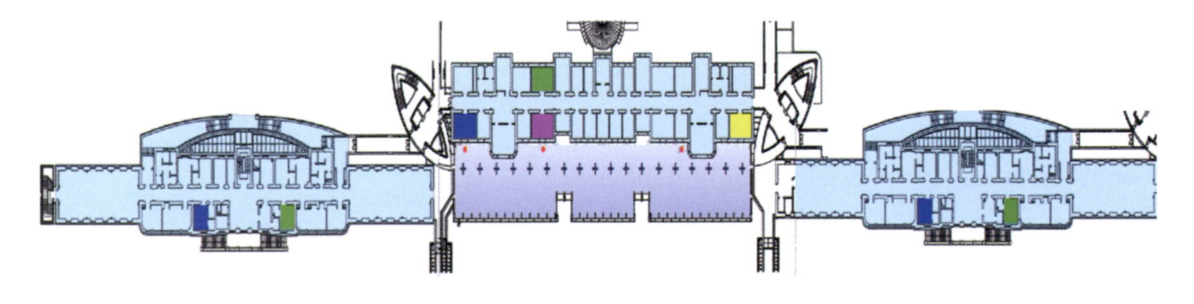

Figure 13.2 Plan of the east–west running terrace showing the additional greenhouse (centre)

Figure 13.3 Replacement timber windows with low-e glazing

Figure 13.4 Refurbished traditional louvre shutters on south elevation

In the end, bearing in mind the nature of the climate (mild Mediterranean) it was decided to leave the walls non-insulated other than a small improvement brought by new external render containing a natural lightweight volcanic aggregate. The resultant U-value is calculated to be 2.0W/°Cm^2.

A new building energy management system (BEMS) was also installed. This not only managed the conventional heating and cooling, but also the night cooling.

Originally, the renewed south-facing windows were to have included an interpane shading system, magnetically controlled from outside the sealed unit. These were finally rejected on the grounds that since the external louvred casement shutters (Figure 13.4) had to be kept for historic reasons, the expense of the additional integral shading was not justified.

The greenhouse

The boldest intervention is the installation of a large greenhouse on the south side of the central pavilion abutting the lower two of the three storeys (Figures 13.5, 13.6 and 13.7). As an energy conserving measure, this strategy may be a little surprising for a warm climate. However, there were two mitigating circumstances – firstly much of the lower glazing is shaded by vegetation, and being in contact with the massive non-insulated wall provides stabilizing thermal mass. However, considerable interest has been

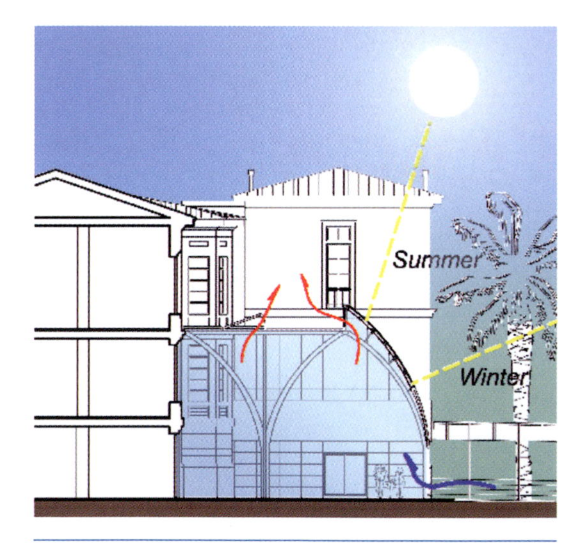

Figure 13.5 Section of greenhouse showing shading and ventilation openings

focused on the risk of overheating in the greenhouse itself, and in the adjacent offices.

The structure is made from laminated timber columns and is a striking addition to the architecture. Its spatial function is as the arrivals foyer and reception area for the whole hospital complex.

Figure 13.6 Interior view of greenhouse showing semi-transparent PV shading

Figure 13.7 Exterior view of greenhouse during construction

Recognizing the potential risk of overheating, the design demonstrates the sound principle of shading and ventilation. In this case, the shading

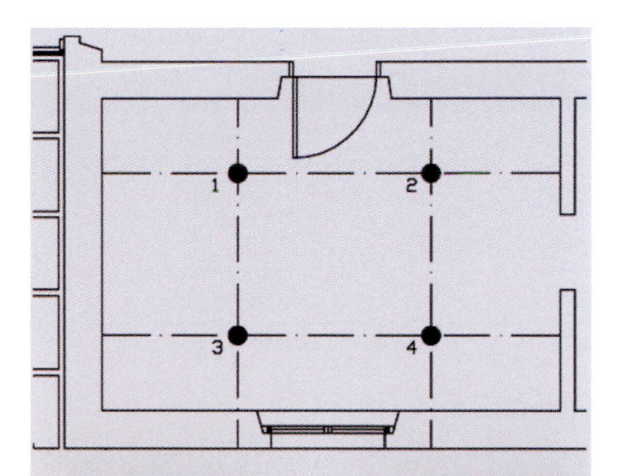

Position	Daylight Factor %
1	1.71
2	1.37
3	2.57
4	2.02
Average	1.91

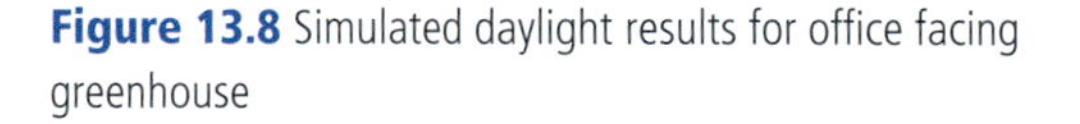

Figure 13.8 Simulated daylight results for office facing greenhouse

is provided by semi-transparent photovoltaic panels. This, and the ventilation provision, are shown in Figure 13.5. The ventilation openings at the top and bottom of the greenhouse are controlled by the BEMS. A further open area can be obtained by leaving the main doors open.

Table 13.1 CO_2 emissions for corrected measured Meyer 2007 data compared with reference case

	kWh	kWh/m²	CO_2 factor	CO_2
Gas reference	1,257,600	120	0.19	238,944
Meyer	876,652	83.7	0.19	166,563
Electricity reference	1,194,720	114	0.44	525,676
Meyer	880,320	84	0.44	387,340

Note: Total CO_2 kg – Reference 764,620; Meyer 553,903.

Daylighting

The impact of the greenhouse on the daylighting of the adjacent office rooms had been tested by simulation (Figure 13.8). The result shows a mean value of 1.91 per cent.

Occupants report a high level of satisfaction with the lighting, 4.0 on a 1–5 scale in the end pavilions, but slightly less, 3.5 in the central pavilion adjacent to the greenhouse (Figure 13.9).

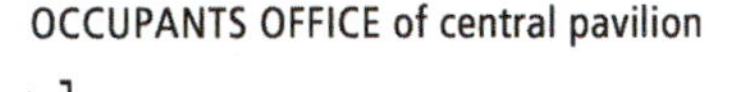

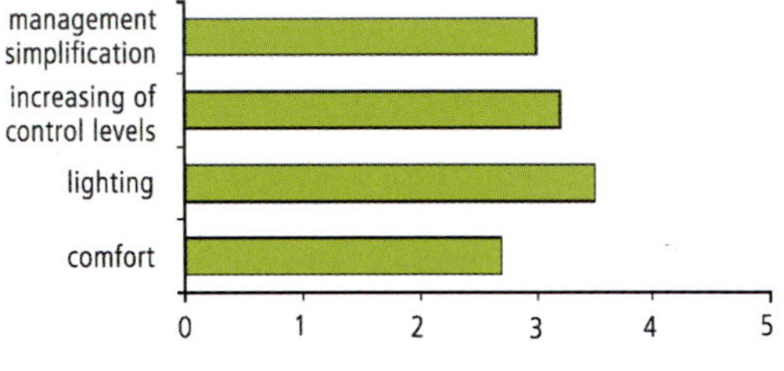

OCCUPANTS OFFICE of lateral pavilion

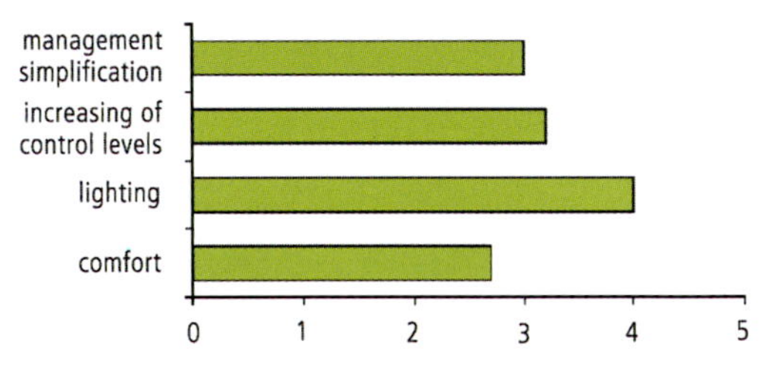

Figure 13.9 Results of user survey for office facing greenhouse (central pavilion) and office facing open air (lateral pavilion). Scale ranges from 0 – very dissatisfied to 5 – very satisfied.

Overall energy performance

The Meyer project is one of those where there has been major change in the function of the building, so there is no comparable data available for energy consumption before the refurbishment. Thus a reference office has been defined.

This shows a 28 per cent reduction in emissions, although it has to be said that the reference values for gas and electricity per m² are very high for a building in the climate of Florence.

Comfort

Questionnaires were issued to occupants in the central wing adjacent to the greenhouse, and in the two lateral wings. The results are shown in Figure 13.10. These are consistent, with the exception of a slightly lower satisfaction for lighting reported in the central wing, which could be due to the presence of the greenhouse. Unfortunately the question does not distinguish between artificial lighting and daylighting.

Comfort, which is probably strongly weighted towards thermal comfort, is a little on the positive side of neutral, which is typical where discomfort is not present, and should therefore be regarded as successful, although since all values are averaged, it is not possible to see if discomfort was reported by some respondents.

It is interesting to note that a positive response is recorded for the level of user control. This indicates that the temperature controllers in each room, providing a simple user-friendly adaptive opportunity, are appreciated.

Appendices

Revival

TECHNICAL**MONOGRAPH** 1

Phase change materials in buildings – Virtual thermal mass

This technical monograph is one of a set produced as part of the 'REVIVAL' project – an EU Energie Programme supported demonstration project of energy efficient and sustainable refurbishment of non-domestic buildings in Europe. The monographs explore some of the main energy and comfort issues which arose during the Design Forums held with each of the six sites. The four monographs are entitled:

- **Thermal mass and phase change materials in buildings**
- Adaptive thermal comfort standards and controls
- Natural ventilation strategies for refurbishment projects
- High performance daylighting

Energy efficient refurbishment must consider the thermal performance of the existing building and how it may be changed for the better (or worse) by the intervention. Phase change materials (PCM) offer a possibility for improving the thermal response of the building by changing the effective thermal mass.

What is thermal mass?

It is the part of the building which stores and releases significant quantities of heat during a typical daily cycle. It is important because it can increase the usefulness of solar gains and internal gains, and reduce the risk of overheating (Figure 1).

Dense materials have more effect than lightweight materials in providing thermal mass, but it is essential that the mass is accessible from the occupied space and/or the space in which the gains are made (Figure 2).

▼ **Figure 1**
The temperature in a lightweight building (low thermal mass) rises above the overheating threshold, whereas that in the heavyweight building does not, although the average daily temperature is the same and no solar gains are available.

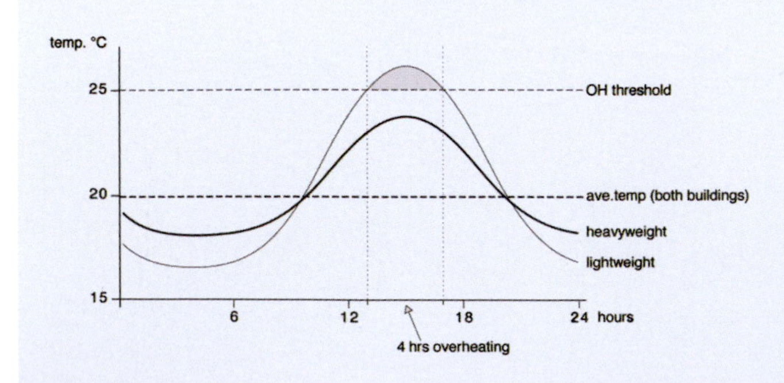

▼ **Figure 2**
Thermal mass must be thermally coupled to the occupied space. It must also have the maximum surface area possible. Thermal mass can be coupled remotely by a mechanically driven air flow.

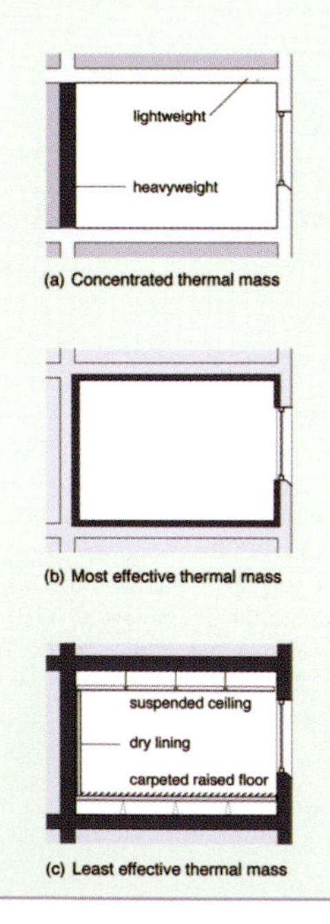

▶ **Figure 3**
Surplus solar gains made at midday are stored in the mass, and become available to offset the heat load later in the day when the outside temperature drops and there is a net heat load.

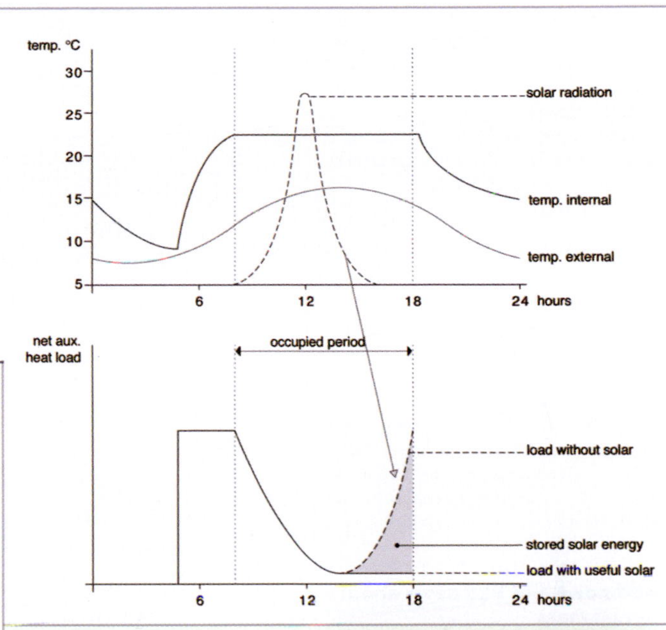

How does it help us?

Winter sunshine may bring ½ kW into the room for every m^2 of south-facing glazing. This may be much greater than the current heating load for the room and in a room with no capacity to store heat, would lead to overheating, and/or the rejection of the heat by opening the window or running the A/C plant. A room with accessible thermal mass would heat up much more slowly because more heat would be absorbed by the mass. When the sun is no longer present and the temperature of the room drops, this heat will be released into the room, delaying the need for auxiliary heating (Figure 3). Thus, thermal mass is normally regarded as an essential constituent of, *solar architecture*.

In summer, the effect is similar, except that the stored heat is probably not required at all, and is taken out of the mass by ventilating the building when the air temperature is lowest, usually at night (Figure 4). The beneficial effect is that it reduces peak daytime temperatures by a much as 3-5 °K.

▶ **Figure 4**
Thermal mass which absorbs unwanted gains in the daytime, has to be 'emptied' of stored heat to be ready to absorb the gains the next day. This is most effectively achieved by high rates of ventilation at night, when the air is coolest.

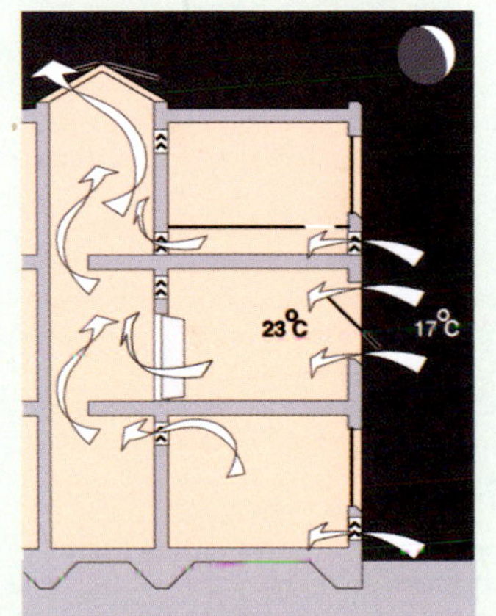

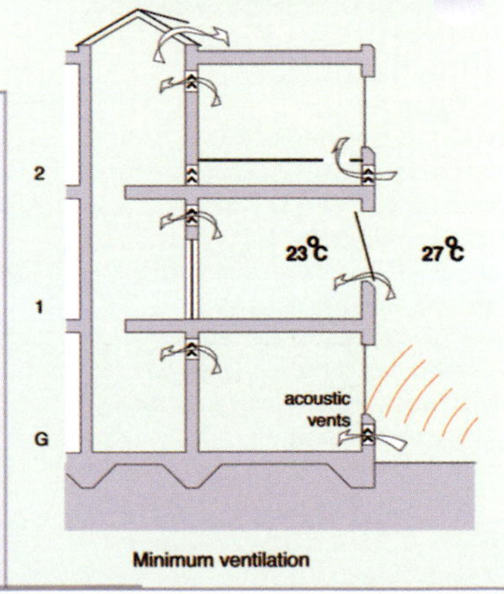

The physics of thermal mass

Thermal mass can be quantified. It is the product of the mass (in kg) and the Specific Heat (in Joules/kg °C). The Specific Heat is the heat required to raise the temperature of one kg of the material by one °C (Figure 5). It follows that the units of thermal mass are Joules/°C.

▼ **Figure 5**
The definition of Specific Heat. The quantity of heat required to raise the temperature of one kilogram of the material by one degree Centigrade.

temp. °C
1Kg
1°C
time
Q = specific heat C

Thermal mass is only effective in storing heat if the heat can flow into it. The deeper into the material, the less heat flows into it (in a given time) and so the less impact it has (Figure 6). The result is that for diurnal (24 hour) temperature cycles, for thermally massive materials such as masonry or concrete, only the first 50mm has much effect. For longer cycles – e.g. a sequence of hot days (or heat wave), deeper thermal mass will slow down the gradual heating up of the building. Only very

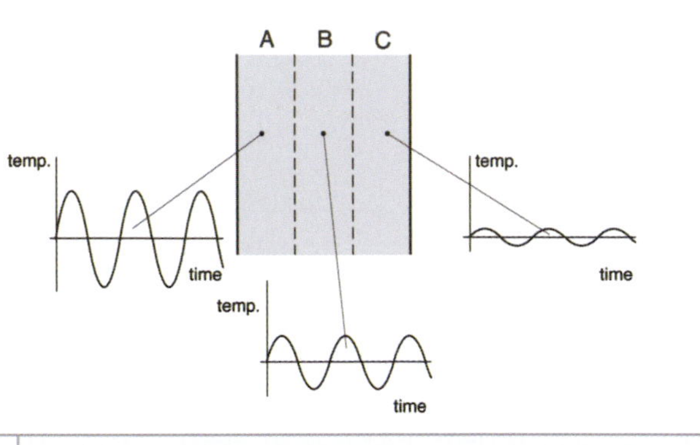

▶ **Figure 6**
When a surface is subjected to a temperature swing heat flows into the mass in the form of a wave whose amplitude diminishes with depth. This means that the deeper the mass, the less heat is stored (per unit volume) for a given temperature swing at the surface, i.e. layer C stores much less heat than layer A.

deep mass, e.g. that which might be found in buildings such as cathedrals and castles, will have a significant effect on a seasonal level – i.e. mean monthly temperatures inside will lag behind the mean monthly temperatures outside. These effects are described as *thermal inertia*.

Good news and bad news about thermal mass

Whilst thermal mass has the benefit of reducing swings in temperature which could lead to overheating, and of increasing the benefit of solar gains in winter, it carries some penalties.

In intermittently occupied buildings (e.g. a primary school) a heavyweight building will require more pre-heating than a lightweight building. Since a large proportion of the heating energy is used at this time, when there are no useful internal gains present, the difference can be quite significant. This is illustrated in Figure 7.

In providing thermal mass that is coupled to the occupied space the acoustics may have to be compromised. Thermally massive surfaces are usually hard and dense, and the resulting acoustic will be noisy and reverberant. In practice, typical office buildings for example, have carpeted floors and suspended ceilings, providing a good acoustic environment, but very poor coupling to the thermal mass in the structure.

What is a phase change material?

The *change of phase* of a material means it changes from a solid to a liquid or from a liquid to a gas. Due to the re-arrangement of the molecules, it involves a considerable amount of energy. We are aware of this when we feel the cooling effect of wetness on the skin, or use a small piece of ice to cool down a drink.

Figure 8 tells this story very well. A fixed mass of water (initially ice) is steadily heated up from

▲ **Figure 8**
This graph demonstrates the thermal effect of phase change. It shows the temperature of a fixed mass of water, initially at –10°C, as it receives a steady input of heat. Note that as soon as phase change commences (melting) heat is absorbed without raising the temperature. The amount of heat required for melting is roughly equivalent to raising the temperature of the same mass 144°K.

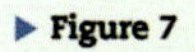
▶ **Figure 7**
Heat loss from the building is proportional to the difference between the inside and outside temperature, integrated over the day. This is the area between the temperature curves for the buildings and the external temperature curve. It is larger for the heavyweight building than the lightweight building.

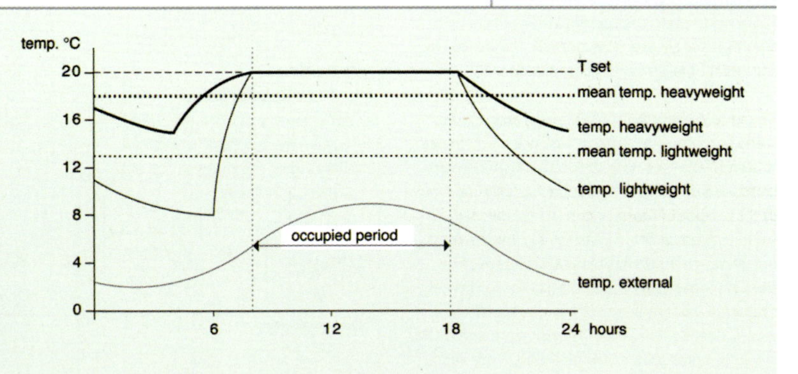

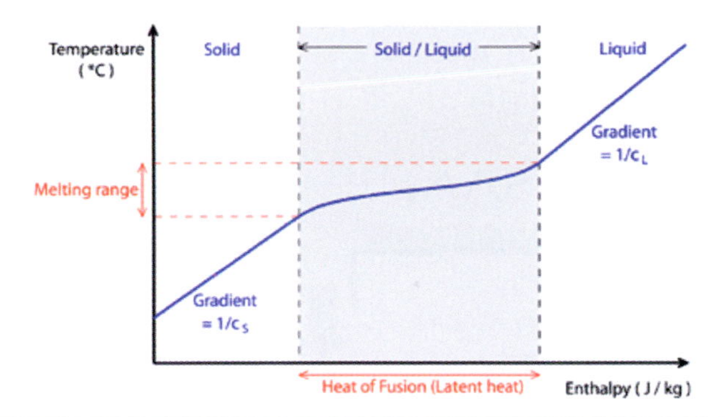

-10°C. Up to 0°C, the ice behaves like normal thermal mass, increasing its temperature as it absorbs heat at a rate of 4.2 kJ/kg °C. As it reaches 0°C it begins to melt. This process absorbs large amounts of heat without increasing in temperature. In fact it absorbs 322 kJ/kg. When all the ice is melted, i.e. the change of phase is complete, the temperature continues to rise. The heat that causes a change in temperature is often referred to as *sensible heat*, because it is sensed by a temperature rise, whilst the heat that changes phase is referred to as *latent heat*.

Now consider the effect of using a material that melted at 24°C instead of 0°C placed on the surface of room in contact with the air. As the room temperature increased above 24°C, heat would flow into the PCM causing it to melt, instead of causing the air to heat up further. If there were sufficient PCM to continue to melt until the peak heat input was past, overheating would have been prevented. It must be pointed out, however, that the room will now cool down more slowly, since as the PCM solidifies, the latent heat absorbed in the melting process will be returned to the room.

In practice, the phase change takes place over a range of temperatures as in Figure 9 (Figure 8 is a theoretical curve). This is partly due to the need for temperature gradients in the PCM in order for the heat to flow, and partly due to the physical chemistry of the materials used.

We can see then that the PCM is behaving as thermal mass, but because of the large amount of heat involved, it is very "concentrated". For example a 10mm layer of PCM material has the effect of about 50mm of concrete. However, it must not be forgotten that the heat absorbing process of normal mass takes place over all ranges of temperature, whereas the PCM is only effective at a specific temperature range.

What are the actual materials used?

There are two classes of materials – waxes and salts. Waxes are chains of polymer molecules (paraffins) and solidify into a crystalline form. The salts are used as a saturated solution in water – solidification involves the water molecules and the salt ions making a crystalline structure. The commonest salt used is Sodium Sulphate ($Na_2SO_4.10H_2O$) or Glaubers salt, but additives are put in to tailor the melting point temperature to the exact requirement of the application.

Good news and bad news about PCM

As we have seen, small amounts of PCM have the effect of large amounts of normal mass. Thus, in a lightweight interior such as our office example (carpeted floor, suspended ceiling), in principle, sachets of the PCM material could be laid on metal trays in the ceiling, converting the interior to "thermally massive". Furthermore, the transition temperature for overheating prevention, would be chosen to be somewhere

▲ **Figure 9**
Realistic melting curve for PCM showing melting temperature range. Enthalpy is the total heat energy; sensible + latent and is equivalent to the time axis in Figure 8

▼ **Figure 10**
Materials used in PCM applications. This graph is presented differently from the melting curves and shows the effective heat capacity (thermal mass) as a function of temperature. The temperature range under the curved part of the graph is the melting range.

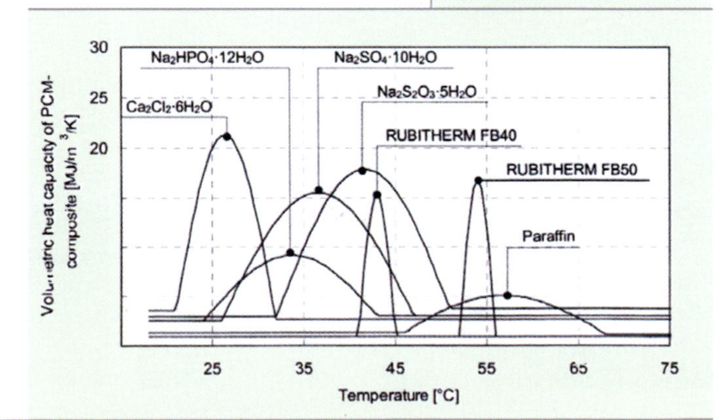

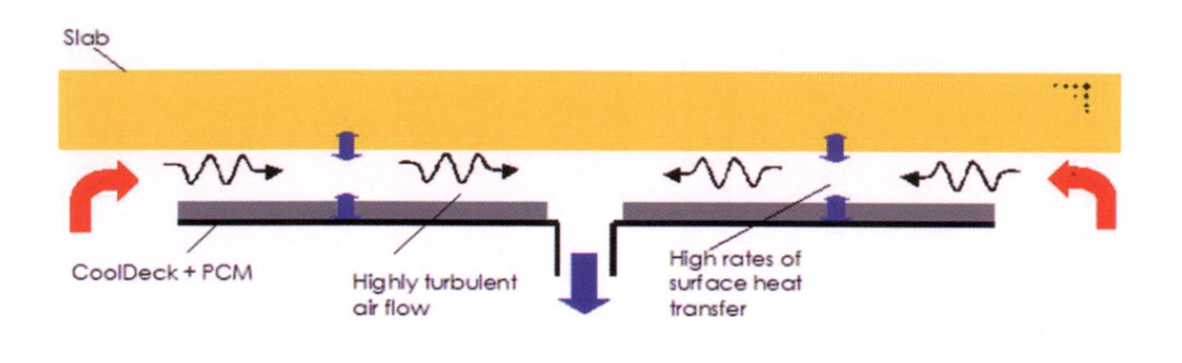

▲ **Figure 11**
The "COOLDECK" system as used at the REVIVAL project in Stevenage. The remote phase change thermal store is coupled to the occupied space by ducted air drawn in from the room via a grille in the suspended ceiling. The air is also brought in contact with the conventional mass of the slab.

in the mid 20ths and thus would not carry the disadvantage of conventional thermal mass during the warm-up period.

The bad news is that PCM is mono-functional and an extra cost, whereas conventional thermal mass – a masonry wall or a concrete floor slab – is already serving an enclosing and/or structural function. Secondly, most PCM materials have a finite lifetime in terms of cycles – some kind of degradation of the material takes place. Thirdly, in spite of the large heat capacity at the melting temperature, due to the relatively poor thermal conductivity of the material, large surface areas must be available to passively[1] couple the material with the space; this may compromise acoustic performance and present other practical difficulties.

The latter problem has prompted designers to adopt hybrid systems involving fan driven airflow to couple the PCM heat sink to the occupied room in the daytime, and to the cool night air at night. This approach carries extra costs, but when compared with conventional A/C, for capital and running costs, is very favourable. It also allows stricter control. This principle is adopted in the REVIVAL project for Stevenage Council offices in the UK, (Figures 11–13).

PCM application design configurations.

The way that the PCM is coupled to the occupied space, i.e. the way that it can absorb and liberate heat to the space, is a critical part of the application.

PCM is available in several different forms but the commonest is in sealed plastic sachets (Figure 13) which are used in conjunction with a support system such as metal trays or ceiling tiles. PCM is also available incorporated into wall coverings – "PCM wallpaper" and plasters. A third approach is to locate the PCM remotely in a "thermal accumulator" (Figure 14), coupling the space via mechanically driven air or water.

1 By natural convection or radiative exchange between the PCM and the room and contents.

▼ **Figure 12**
Metal trays containing PCM attached to underside of concrete slab, coupled with room air by forced convection in COOLDECK system used at Stevenage

▼ **Figure 13**
PCM materials in sachets as used in Cooldeck system at Stevenage

▶ **Figure 14**
PCM thermal store at the Bristol Science Centre, by Chris Wilkinson Architects and Arup.

Bibliography

Kovach, E. G. (1978), *Thermal Energy Storage*, Oxford, Pergamon Press.
Ure, Z. (1998), *Eutectic Thermal Energy Storage Concept*, Published as part of IEA Annex 10, Phase Change Materials and Chemical Reactions for Thermal Energy Storage, 1st Workshop, 16-17 April 1998, Adana, Turkey.
Stacey, M. (October 2003), "Theme: Cladding" in *AJ Focus*, London, EMAP Communications.
Papers published as part of: IEA, ECES IA Annex 17, *Advanced thermal energy storage through phase change materials and chemical reactions – feasibility studies and demonstration projects. 3rd Workshop, 1-2 October 2002, Tokyo:*
Ip, K. et al. (2002), *Thermal Storage for Sustainable Dwellings*, University of Brighton, UK.
Kitano, H. et al. (2002), *Simulation Approach for Applicability of Phase Change Composites in Architectural Field*, Mie University, Japan.
Kuroki, T. et al. (2002), *Application of Phase Change Material to Passive Cooling of Apartment House*, Takenaka Corporation Research and Development Institute, Japan.
Lamberg, P. et al. (2002), *The Effects on Indoor Comfort when using Phase Change Materials with Building Concrete Products*, Helsinki University of Technology, Finland.
Mehling, H. et al. (2002), *News on the Application of PCMs for Heating and Cooling Buildings*, ZAE Bayern, Germany.

Summary conclusions

PCM offers a way of increasing the effective thermal mass of a building, requiring less space than the equivalent 'conventional' mass. Since it can be retro-installed it has special value in refurbishment projects.

PCM installations can be made more effective by mechanically coupling the material with the space via fan-driven airflow; this also allows closer control which improves system efficiency. The Coefficient of Performance[2] of such a system at Stevenage is claimed to be 11:1 compared to about 3:1 for a typical A/C system.

Due to large surface area requirements, it is inherently a 'low capacity' system when used passively. As with all passive and hybrid systems 'prevention is better than cure'. In other words minimise the cooling loads by appropriate shading and internal gains reduction before specifying the PCM system.

In its heat storage role, PCM offers a way to reduce global energy consumption, only if it is used in conjunction with ambient heat sources or sinks – e.g. solar energy or the cool night air. Ice storage used in A/C systems, or PCM used in off-peak heat storage systems only delay the demand for conventional energy to a period when it is cheaper; it does not reduce the total energy use.

Prepared by Nick Baker the Martin Centre University of Cambridge

Acknowledgements

I am grateful to Andrew Macintosh for the use of Figures 9, 10, 11, and 14.

2 The ratio of heat displaced by the system to the electrical power to drive it

Revival

TECHNICAL**MONOGRAPH** 2

Adaptive thermal comfort and controls for building refurbishment

This technical monograph is one of a set produced as part of the 'REVIVAL' project – an EU Energie Programme supported demonstration project of energy efficient and sustainable refurbishment of non-domestic buildings in Europe. The monographs explore some of the main energy and comfort issues which arose during the Design Forums held with each of the six sites. The four monographs are entitled:

- Thermal mass and phase change materials in buildings
- **Adaptive thermal comfort standards and controls**
- Natural ventilation strategies for refurbishment projects
- High performance daylighting

Improving energy performance and comfort

Refurbishment almost always involves the installation of new plant and controls. This in itself is often an energy-saving strategy since, not only is heat (or cooling) produced more efficiently in modern plant, but improved controls also result in far higher utilisation – i.e. a reduction in wasted heating or cooling.

Another motivation for plant replacement could be to improve comfort where for various reasons existing plant is inadequate. This may be due to the degradation of the system and controls, or due to gradual increase in heating (or cooling) demand due to increase in the total built area. Caution is needed here when specifying the capacity of plant, since other measures may result in significantly reduced loads – hitherto inadequate plant may now be able to cope. Even if for reasons of technical efficiency the plant is to be replaced, then it can probably be much smaller.

▼ **Figure 1**
Ad hoc additions to existing air conditioning plant without improvements to the fabric, together with old and degraded heating systems can often lead to very low efficiencies.

Thermal Comfort Standards

Once the decision to replace plant and controls has been made, then the question of comfort standards arises. There tends to be a presumption that air-conditioned buildings always provide a superior level of comfort. However, it is often found that satisfaction levels are quite high in naturally ventilated 'poorly serviced' buildings – for example the database of Building Use Studies shows that occupants are about equally satisfied in air-conditioned and non air-conditioned buildings.

There will be some cases where new plant will lead to major improvements in comfort conditions, but may lead to an increase in energy consumption of the refurbished building, in spite of other improvements. This would be the case, for example, in warm climates where the original building may have had no heating or cooling system at all (Figure 2). In these cases, the added value of improved comfort has to be recognised in assessing the performance of the improved building. It may be appropriate to make a comparison of the refurbished building with the predicted energy cost of the old building if it were serviced to provide the same comfort standards.

In the case of refurbishment, there may be qualities in the original building that are lost. For example, openable windows are often removed when re-glazing or double skinning,

and views maybe compromised by the application of shading devices. There is some evidence that building features such as these make occupants more tolerant of temperature swings. And it may be just as satisfactory to allow some drift in temperature conditions as in the original building, rather than implement a close temperature control strategy at considerable energy cost.

In order to achieve an informed solution to this question, it is necessary to understand the basic principles of conventional comfort standards and the more recent understanding of how *adaptive behaviour* can also influence overall satisfaction. Adaptive behaviour is where the occupant interacts with the environment in order to improve it, or changes their own state to improve their heat balance. This is an elaborate way of saying for example – "opens the window or takes their jacket off". There are also psychological factors that may influence the occupant's response to a given thermal condition. Commonplace though these responses are, conventional comfort theory has virtually ignored this aspect; in particular the extent to which it is influenced by building design.

Conventional Comfort Theory and Heat Balance

Figure 3 illustrates the energy balance of the human body – or indeed any warm-blooded animal. Chemical energy is taken in the form of food, converted to mechanical energy and heat by the process of metabolism. The heat is an essential 'bi-product' in this conversion. This 'waste' heat production is found in other instances of energy conversion – e.g. the engine of a motorcar or at a power station.

However, in warm-blooded creatures nature has put this heat to good use, by maintaining the body at a steady temperature above that of the surrounding environment. This allows chemical processes such as enzyme action, to be optimised and independent of external conditions. This gave a huge advantage to warm-blooded creatures compared with cold-blooded animals. The problem is that this steady temperature has to be controlled to within +/- about 1°K. In humans this is achieved by a series of responses some of which are physiological (e.g. shivering, sweating) and some of which are behavioural (e.g. putting more clothes on).

▲ **Figure 2**
The refurbished Meyer Hospital administration block is to be air-conditioned, and although it will be a highly efficient system, it will lead to an increase in energy consumption since the original building had no heating or cooling services.

▶ **Figure 3**
Metabolic heat is generated as a bi-product of the conversion of food energy to useful work.

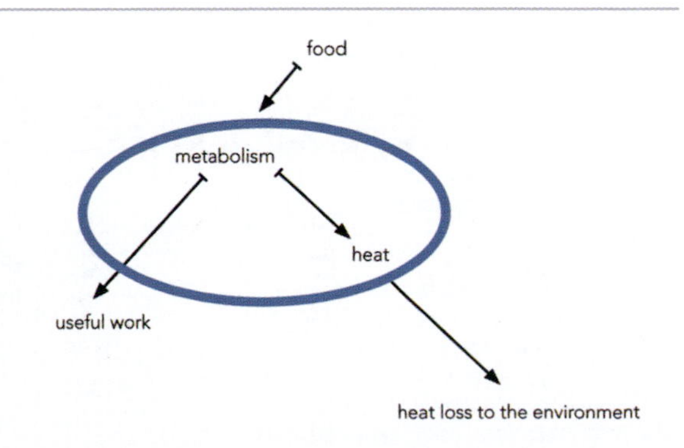

Heat balance is the key to the process – if the heat loss to the environment is less than the metabolic heat gain, the person will go on getting hotter and hotter, and if heat loss is more than the heat gain, the person will get colder and colder. In both cases the ultimate result will be death. Feelings of discomfort are to give us early warning of this serious outcome.

Conventional comfort theory, as pioneered by P.O. Fanger, focuses on the need to provide this heat balance at all times by controlling the environmental temperature. In order to establish these 'ideal' conditions, subjects were placed in a climate chamber where the temperature was varied until the consensus response (the mean vote) was of zero thermal sensation. This was correlated with various parameters, such as metabolic activity level and the total heat loss (calculated from the person's temperature and clothing insulation level).

It is easy to see that this produces a universal figure – an office worker dressed in a suit operating a computer in Stockholm would need the same thermal environment as one similarly dressed in Calcutta. The internationalisation of standards through bodies such as ISO and ASHRAE has resulted in the global growth of the air-conditioning industry, and has contributed to the ever-increasing use of fossil fuel.

Adaptive Comfort Theory

The reality is however, that millions of people are satisfied with environments that fall a long way outside the limits prescribed by Fanger. Humphreys in his 'survey of surveys' carried out in 1974, showed that using the data from many surveys taken all over the world, the neutral temperature at which people reported satisfaction bore a positive relationship to the outside monthly mean temperature as shown in Figure 4. How can we explain this?

It's not that the physics of Fanger's analysis is wrong, rather it is the extrapolation of the subject's responses, from the steady state climate chambers, to normal living and working conditions that is inappropriate. This is because in real life, the body is not in steady conditions, and has at its disposal the use of many behavioural mechanisms, as well as physiological ones, to achieve a long-term thermal balance.

Furthermore, psychological factors influence our *interpretation* of non-neutral thermal sensations – for example, if we are a bit chilly, the warm spring sun can be a delight; if we are already sweating from insufferable heat, that same thermal sensation would certainly not be.

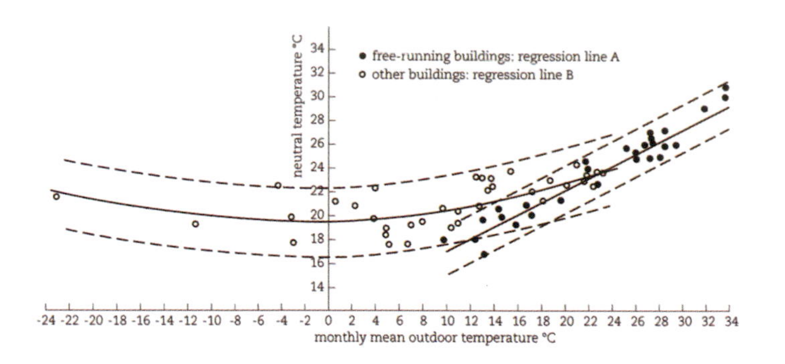

We are sensitive to the degree of risk of long-term heat imbalance. The momentary feeling of chill when we run out of a heated building to fetch something from the car, is interpreted very differently from feeling only slightly chilly in the middle of an open-plan office where we have no control over the temperature.

Adaptive actions, and the interpretation of the long term situation, are then, a key factor in satisfaction with the environment. It follows that a vital quality of an environment is the opportunity to make adaptive actions. This is illustrated in Figure 5 which shows an acceptable 'comfort zone' within hypothetical varying stimulus, (which could be temperature). Note that with good adaptive opportunity (a) the zone widens to include the extreme values, that in the intermediate case (b) the subject is stressed only when the adaptive zone is exceeded, whilst where there is no adaptive opportunity (c), any departure from neutrality results in stress.

It is important to note that conditions of zero adaptive opportunity are those prevailing in the climate chambers. This accounts for the far more demanding conditions for comfort resulting from climate chamber work.

▲ Figure 4

The open circles are for surveys carried out in buildings that are air-conditioned, the solid circles are for free-running buildings. It shows that the comfort temperature bares a linear relationship with the monthly mean outdoor temperature. Fanger's equation would conclude that the comfort temperature would follow a horizontal line. The increasing comfort temperature is explained by the effect of adaptive actions by the occupants. It is not a physiological adaptation as often suggested. Humphreys derived the equation

$T_n = 11.9 + 0.534\, T_m$ +/- 2.5K
where T_n is the neutral temperature and T_m is the mean monthly temperature

▼ Figure 5

The neutral zone is, in effect, extended by adaptive opportunity. In the case of poor adaptive opportunity, the swing in the stimulus (e.g. temperature) exceeds a value with which the adaptive opportunity can cope leading to stress. In the case of no adaptive opportunity, any departure from the conventional comfort zone leads to stress. This is the situation in the climate chamber.

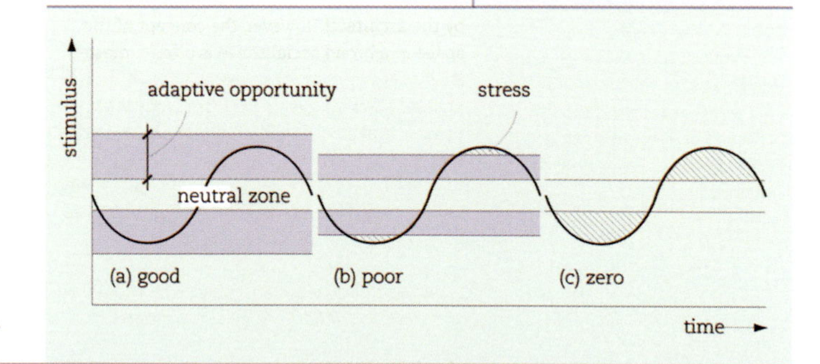

▲ **Figure 6**
Office environments in Portugal showing good (left) and poor (right) levels of adaptive opportunity

Application of adaptive comfort theory

What does this mean for designers and specifiers? How do we provide an environment with good adaptive opportunity?
We can identify three kinds of adaptive opportunity:

- Personal
- Building
- Systems and controls

Personal refers to the freedom of an individual to take actions that affect their own heat balance. Some of these are almost involuntary, such as the way one sits or moves. An important adaptive opportunity relates to clothing – the freedom of an individual to choose a clothing ensemble and/or to make adjustments that alter its insulation value.

For example the insulation value (measured in Clo units) for a pair of shorts and tee-shirt is 0.3 whilst that of a business suit is 1.0. By adopting the former, the comfort temperature will increase by about 3°K – this clearly may be sufficient to avoid air-conditioning with significant energy implications. There is no doubt that the ubiquitous business suit has been a major stimulus to the growth of air-conditioning in office buildings world-wide, even in temperate climates.

Dress code is not an area normally covered by the architect! However, the concept of the space in a broad social sense is often a matter for discussion in design meetings and it could be that informed input from the architect or engineer might influence the client in this respect.

Access to cold or hot drinks is also an adaptive opportunity that can have a significant effect.

Building refers to more familiar elements such as openable windows, occupant controllable blinds, flexible furniture layout and spatial freedom. Due to the variation in personal requirements, it is easier to satisfy occupants in cellular buildings than in open-plan.

It is also found that people are much more tolerant of variation in environmental conditions (for example temperature), if they are close to a window. There is also growing evidence that the provision of good outdoor views affects the overall well-being of occupants, which in turn influences their response to internal conditions in a positive way.

These issues are more familiar territory for the architect and can effect the design of the building from its most fundamental massing and planning, to the detailed specification of elements such as blinds and furniture.

Systems and controls refer to the provision of mechanical systems such as heating, lighting and ventilation, and the access of the occupant to their control. Because mechanical systems offer the possibility of high levels of automatic control, and with IT intelligence, it is often believed that this is preferable to manual control. However, it is now widely accepted, that local personal[1] manual control should be provided wherever possible.

This is not to say that intelligent controls have no role. Ideally, systems should have a *caretaker function*, i.e. they allow personal, manual control, but after a period of time they return the status of the controlled service back to a low-energy standby.

[1] 'Personal' implies a control which ideally would influence a single occupant. This has been provided in some hi-tech workstations, and is the norm in cars and aeroplanes. In buildings it is less common and may inhibit flexibility in room layout. However, controls should be as local as reasonably possible – the fewer occupants affected, the more the control will be used, and appreciated.

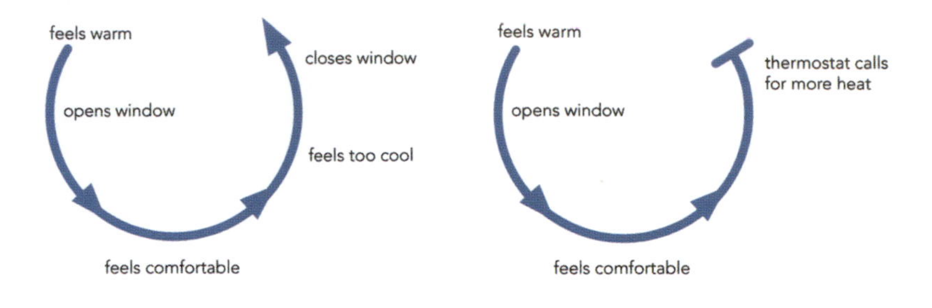

▲ **Figure 7**
The control feedback loop is broken by the intervention of the thermostat

▼ **Figure.8**
Micro-switches sense the opening of the large windows onto the balcony and switch off the air-conditioning. This prevents the feedback loop from being broken and prevents wasted energy

K.A.T. HOSPITAL
SOUTH EAST FACADE

Feedback loops

For example, in the case of heating controls, the normal setting might be in the lower half of the 'comfort zone'. However, an occupant might temporarily require more heat and adjust the setting higher. After a period of higher temperature, the control system would gradually return the temperature to its standby value.

This approach is particularly valid where feedback is weak. A good example of this is lighting. The occupant arrives in the morning in poor daylight and switches on the lights. Later during the day, the sky brightens and daylight alone is sufficient. However, the added illuminance of the artificial light is insufficient to make the room *uncomfortably bright*. Thus there is no incentive for the occupant to turn the lights off. The caretaker control would test if there was sufficient daylight available and then dim out the artificial lights.

In other circumstances, automatic controls 'sabotage' the natural feedback loop.

Consider the case shown in Figure 7 where an occupant, denied access to local temperature controls, opens a window. Instead of the room temperature dropping, thereby prompting the occupant to close the window, the thermostat calls for more heat. This not only frustrates the occupants intentions, but is also very wasteful of energy. It is this kind of conflict between the user and automatic control that has led engineers to develop hard line attitudes to occupant control.

What is needed, is a control system in which occupant action is not ignored or prevented, but becomes part of the system. In the example above, by simply disabling the thermostat when the window is opened, the natural feedback loop is restored. Further energy saving could be achieved by setting the heat demand to a lower or even zero level. This approach is to be incorporated in the KAT hospital refurbishment (Figure 8) where a micro-switch in the sliding doors to the balcony, switches off the room air-conditioning.

Intuitive interface design

The visual and ergonomic design of the control interface is important. Much attention is paid to this in other design areas such as motorcars. In the IT area, the issue of the degree to which the use of an interface is intuitive has become important in software design. Rather less attention has shown to this in building controls.

Controls to passive elements, such as blinds or windows, are naturally intuitive because the user understands the mechanics and interacts directly with the element; opening a window obviously connects the inside with the outside and therefore the outside air. This is often not the case with controls to mechanical systems. If anything, the situation has worsened with

Positive adaptive attributes:
Relaxed dress code
Occupant mobility – access to hot/cold drinks
Openable windows
Adjustable blinds
Desk fan or locally controlled ceiling fan
Local heating/cooling controls
Workstation/furniture flexibility
Shallow plan (minimizing distance from windows)
Cellular rooms (reduces mutual disturbance)
Surface finishes appropriate to visual task
Daylight and task lighting backup
Good views (external and internal)
Transitional spaces (verandahs, atria etc)
Good access to outside areas

Negative adaptive attributes:
Uniformity of physical environment (temperature, lighting, colour)
Deep plan, reduced access to perimeter
Dense occupation with restricted workstation options
Sealed windows
Views obstructed by fixed shading devices
Central mechanical services control

growth of IT and digital control – multi-function buttons and touch-screens are a long way from a control knob with an arrow and HOT – COLD scale.

Finally, even if the control interface is intuitive, it must also be ergonomically satisfactory. A window catch that can only be reached by standing on the desk, or a radiator thermostat valve at floor level behind a filing cabinet, will not be used frequently. Another related matter is the importance of providing modulation control where required. For example, an opening window that is either closed of wide open, or a blind that cannot be partially deployed, is unsatisfactory, and will be used far less. The detailed design of control systems is beyond the scope of this monograph, but the four key principles are

- anticipating occupant interaction
- maintaining feedback to the occupant
- caretaker function
- intuitive and ergonomic control interfaces

How effective is the provision of adaptive opportunity?

The impact of providing adaptive opportunity is diffuse. There seems to be a general improvement in well-being indicated by higher scores in overall satisfaction in post occupancy evaluations (POE). This carries benefits to the building operator of higher productivity and low absenteeism rates.

However, it would be useful to be able to quantify the effects on specific environmental parameters such as temperature, i.e. by how much does the provision of adaptive opportunity affect the acceptable temperature limits. If this question could be answered, then conventional and internationally accepted temperature standards, derived from non-adaptive comfort theory, could be modified in response to the degree of adaptive opportunity.

Relatively little work has been done in this area although in a pilot study for CIBSE values of 'adaptive increments' for various adaptive opportunity attributes were tabulated, as shown below. Though the temperature increment is easy to specify quantitatively, the adaptive attribute has to rely on verbal description.

The adaptive opportunity table above can be used as follows:

1. look up the conventional comfort standard
2. select the adaptive opportunities that are present.
3. judge the degree to which the opportunity is present and modify the increment/decrement accordingly.
4. sum the modified increments/decrements and use this figure to extend the upper comfort limit.

Adaptive opportunity	Comfort temperature **decrement/increment** on standard comfort zone width (+/- 2.5°K from neutral)
Personal	
• free dress code	
• – 0.3 Clo	+ 2.5
• – 1.2 Clo	- 1.5
• Non-upholstered chair	+ 0.5
• Access to cold/hot drinks	+ 0.75 (- 0.5)
• Metabolic rate and posture	+ 1.0 (- 0.5)
Building	
• Desk fan or ceiling fan	+ 2.5
• Openable window	+ 1.5
• Operable blinds	+ 1.0
• Spatial variation	+ 1.25

Metabolic rate and Posture describe the tendency for people when overheated to move more slowly and efficiently, thereby lowering their metabolic rate. People also sit in a more extended posture, exposing a larger surface area than when under heated.

Spatial variation describes the ability of an occupant to seek out the position that is more comfortable than the room average. This could involve a small movement into a moving air stream or a better-shaded part of the room.

There are more opportunities for raising the upper temperature limit than for lowering the lower limit. It is almost universal to provide mechanical heating in cool climates although in some areas of southern Europe heating equipment is not installed. The most noticeable effect is the raising of upper limits, and this may permit the avoidance of air-conditioning.

Example. An architect's office in Athens has openable shaded windows, a cold drinks machine and a free dress code (occupants often wear shorts and tee-shirts). Seating is on open mesh steel chairs.

All adaptive opportunities are available except 5 and 6. Opportunity no.7 has the problem that when the louvre blinds are deployed they inhibit the ventilation and view so we will reduce that to 0.5. This gives a total increment of 8°K.

Taking a 'normal' neutral temperature for offices as 21 +/- 2.5 °C gives an upper 'conventional' limit of 23.5 °C. Adding the increment of 8°K gives us an absolute maximum of 31.5 °C.

In a study, actually carried out in an office in Athens during the PASCOOL project, it was found that the average satisfaction reported was 83% for a spatial mean room temperature of 30.5°C.

Summary Conclusions

There is now widespread support for the adaptive thermal comfort model. Both ASHRAE and CIBSE have recognised the adaptive model, and ISO standards are becoming more flexible in response to data from many EU funded research activities.

It remains difficult to quantify the benefits and the adaptive attributes of a design. This means that it is all too easy for the engineer when faced with a client's need for assurance, to resort to a conventional performance specification, met exclusively by an engineered solution.

However, good adaptive design can emerge from a well-integrated design team, where client, architect and engineer share the problem and approach the design solution together.

Refurbishment offers both constraints and opportunities. Whilst major re-modelling maybe economically impossible, many European buildings of 30 – 50 years old were designed in an era when relatively low servicing was the norm. This often means that shallow plans and generous floor-to-ceiling heights prevailed. Although detailed design, along with 'building abuse' in the form of alterations and 'improvements', may have rendered them currently poor performers, there remains a potential for good passive rehabilitation, which is usually closely compatible with good adaptive design.

The benefits of avoiding heavy servicing in refurbishment projects are considerable – apart from the reduction in capital costs, maintenance and energy, fitting ductwork and plant into existing buildings is often difficult and expensive in terms of space.

Developments in IT can also make a contribution in good adaptive design. Wireless controllers, occupancy detectors and local zonal control all help to give back "power to the people".

Prepared by Nick Baker the Martin Centre University of Cambridge

Revival

TECHNICALMONOGRAPH 3

Natural ventilation strategies for refurbishment projects

This technical monograph is one of a set produced as part of the 'REVIVAL' project – an EU Energie Programme supported demonstration project of energy efficient and sustainable refurbishment of non-domestic buildings in Europe. The monographs explore some of the main energy and comfort issues which arose during the Design Forums held with each of the six sites. The four monographs are entitled:

- Thermal mass and phase change materials in buildings
- Adaptive thermal comfort standards and controls
- **Natural ventilation strategies for refurbishment projects**
- High performance daylighting

Refurbishment projects often involve changes to the ventilation strategy. Commonly this will include the reduction of uncontrolled infiltration, usually as part of a general upgrading of the envelope. The reduced infiltration will reduce heating loads, but may result in indoor air quality (IAQ) dropping below acceptable standards. This will often prompt the installation of mechanical systems as part of the refurbishment package.

▶ Figure 1
The Albatross building in Den Helder has an advanced passive ventilation system incorporated in a secondary glazed skin.

Furthermore, mechanical systems may often be installed to combat overheating, the result of poor design features such as large areas of unshaded glazing. An integrated approach to the refurbishment specification may allow natural ventilation, with the following advantages:

- Save electrical fan-power
- Reduce plant costs and maintenance
- Save space
- Improve health and well-being of occupants

Can we avoid mechanical ventilation?

Natural ventilation is becoming an attractive alternative to mechanical ventilation or full air-conditioning in new and re-furbished buildings. It is of course, a re-discovered technology, since before the 20th century all buildings were naturally ventilated. Furthermore, almost all residential buildings, and many non-residential buildings still rely on natural ventilation. Why the big issue then?

There are two important factors that have changed the situation. Firstly buildings have got much bigger and more complex, and secondly, the influence of energy conservation has required that wasteful over-ventilation is avoided.

Traditionally buildings had leaky envelopes and this often provided a 'failsafe' for air quality. But with the growing trend to make air-tight envelopes to save energy, ventilation has to be designed more precisely. Three different purposes for ventilation can be identified and each needs to be considered separately although their provision may be by common elements:

Ventilation	Purpose	Ventilation rate
Minimal ventilation	To maintain air quality	Typical winter case 0.75 – 1.5 ac/h
Space cooling	To vent unwanted heat	Typical summer case 2 – 12 ac/h
Physiological cooling	To provide direct air movement to occupants	Typical summer case 0.5 – 1.5 m/s

These purposes make very different demands on the building. In winter, the problem is to exchange just enough air to maintain sufficient air quality. "Build Tight – Ventilate Right" implies that the ideal is to have an airtight envelope with purpose made controllable ventilation openings, positioned to give the best mixing and minimise discomfort from cold draughts. Openings may be windows, or closable grilles or trickle vents. If they are windows only, they must be able to be set at a very small opening area.

The driving forces of natural ventilation

- Wind generates pressure differences across the building which cause air to flow through openings in the building envelope, figure 2.
- Temperature difference between inside and outside causes a vertical pressure gradient which causes air to flow vertically (upwards if building is warmer than outside). This is known as buoyancy flow or more commonly as the stack effect, figure 3.

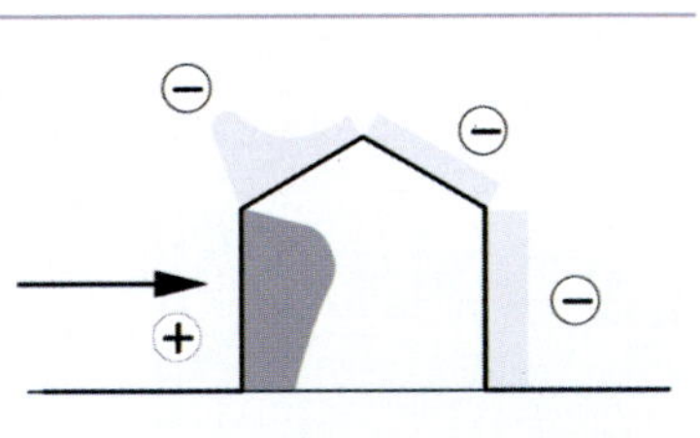

▶ **Figure 2**
Wind pressure distribution. Airflow takes place between openings at different pressures.

▼ **Figure 3**
Temperature difference between inside and outside creates a pressure difference across the envelope driving airflow in through openings at the base and out the upper part of the building.

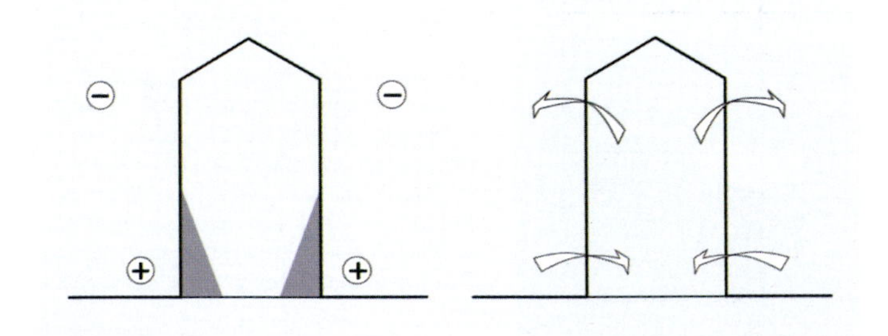

The control of ventilation rate

For much of the time both of these forces are present simultaneously. The problem is that both are highly variable. Overall air change rate on a cold windy day will be many times that on a warm calm day, unless the openings respond to change the flow resistance, figure 4.

Rules for winter ventilation

- Openings should be small and controllable to account for different wind strengths and temperature differences.
- Openings should be high up in the external wall of the room to encourage mixing and minimise draughts.

If only fixed trickle vents are provided, they have to be large enough to cope with the worst conditions – i.e. no wind and small temperature differences. This will lead to over-ventilation in conditions of significant wind and large temperature differences, and hence wasted energy, for much of the time. Automatic vents are now becoming available, which close up as the pressure differences get greater, thereby stabilising the air-flow rate. Or active ventilators under BMS control can be modulated in response to temperature and wind speed, or indoor air quality (IAQ).

Rules for summer ventilation

- Openings should be large and easily controllable (good access to handles, stays, locks etc)
- Openings should be well distributed

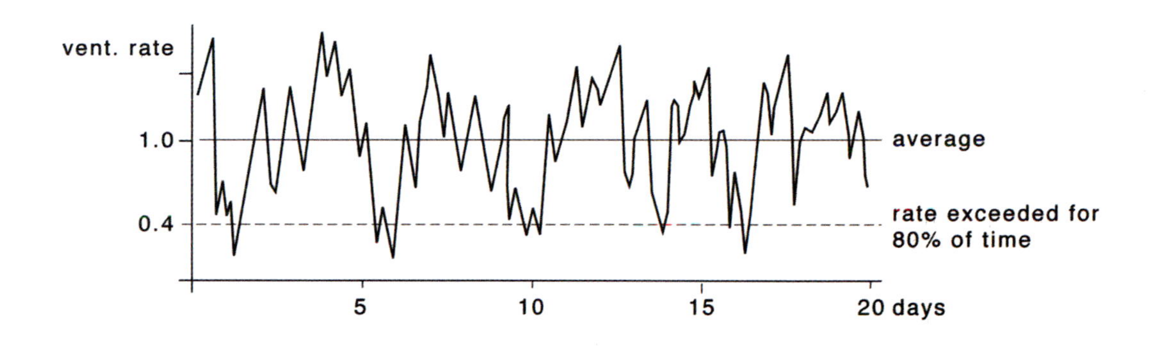

▲ **Figure 4**
Due to the variability of the driving forces, fixed openings sized to provide sufficient minimum ventilation will result in over ventilation (and energy waste) when driving forces are large.

vertically and horizontally to encourage flow between parts of the façade at different pressures (see figures 6 and 7).

- Consideration must be given to how the incoming air will affect the occupants.
- Shading devices must not block summer ventilation openings (see Tech. Mon. 4)
- Openings may require special design features to reduce transmitted noise.

In both cases, consideration must be given to the distribution of fresh air within the space. Different distributions of openings allow different depths of floor plan to be naturally ventilated. The following table (figure 5) gives a rule of thumb.

Note that the depth of effective ventilation is dependent on floor to ceiling height. The removal of a suspended ceiling may represent a refurbishment opportunity for improving natural ventilation (and daylighting), provided other problems such as servicing and acoustics can be solved.

▲ **Figure 6**
Wind generates complex pressure distributions on buildings, particularly in urban environments. This assists ventilation, provided that openings are well distributed and flow paths within the building are available.

▼ **Figure 5**
Depth of effective natural ventilation in rooms from side openings

	Single-sided		Double sided (cross ventilation)
	Single opening	Multiple openings well distributed vertically and horizontally	
Depth of floor in units of floor to ceiling height h	2h	3h	6h

▼ **Figure 7**
For a given total area, ventilation is improved when openings are well distributed vertically and horizontally. This is because air flows at different pressure between openings. It also leads to a better distribution within the room.

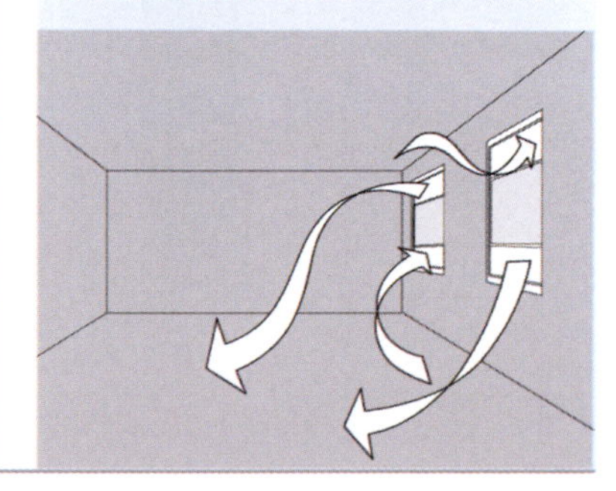

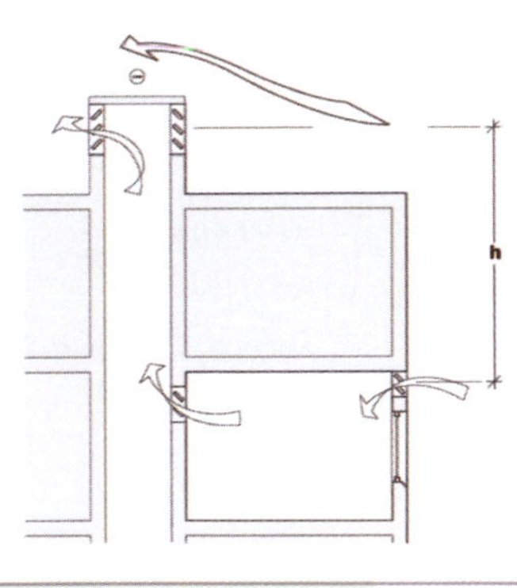

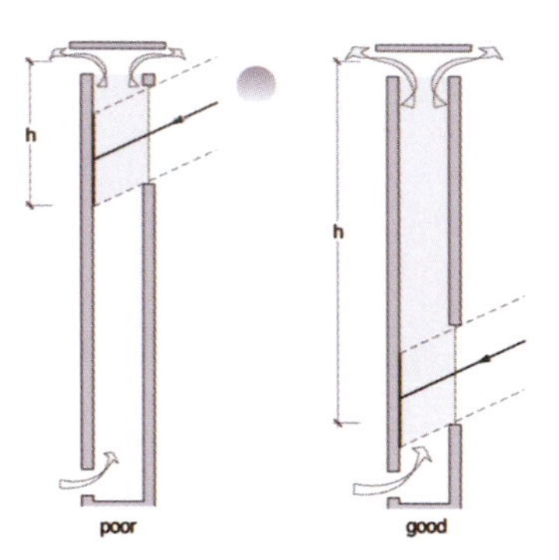

▲ **Figure 8**
Stacks can ventilate a deep plan building; wind and buoyancy forces both create negative pressure causing fresh air to flow in from the perimeter

Advanced ventilation techniques

Ducts and chimneys

So far this guidance has considered ventilation provided by openings in the facades of the building, and usually doubling up as windows. This imposes some limitations on plan depth and layout. To get sufficient quantities of air in and out of an existing deep-plan space, ducts and/or chimneys may be provided, figure 8.

Large vertical ducts can generate larger airflows, by the so-called 'stack effect', than can be obtained in a single room by side openings, because of their greater height. Furthermore, when the wind does blow, there is always a negative pressure across the top of a building; thus the wind driven and stack driven flows complement each other. In order to keep flow resistance low, the cross-section must be quite large, typically between 2% and 5% of the floor area they are serving, assuming a similar area of inlet is available at the perimeter.

This may be difficult to provide in an existing building, although use of existing light wells or redundant lift shafts might be a solution.

▲ **Figure 9**
Solar chimneys use solar gains to heat the air column. However, the driving force is dependent on the height of the warm air column and the average temperature difference. It follows that it is no good heating up the air at the top only.

The performance of the stack can be enhanced by heating the air in the stack by solar energy, but the air must be heated from the bottom, not just as it leaves the stack, since it is the height of the warm column of air that drives the flow, figure 9. These elements are often referred to as solar chimneys; they may have problems in cold and sunless weather when the poorly insulated glass, cools the air and generates reverse flow. Because they need to be heated for the height of the column, they cannot be located in the centre of a deep plan building, unless the solar part protrudes a long way above the roof level.

▼ **Figure 10**
Night ventilation can reduce daytime temperatures by as much as 4°K. However it only works where there is thermal mass available internally, and high rates of nightime ventilation.

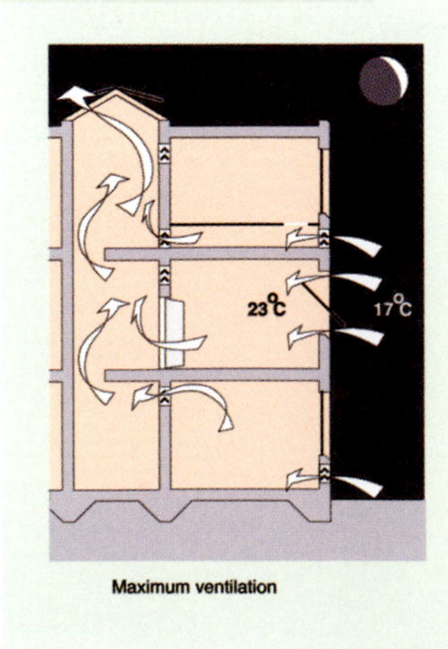

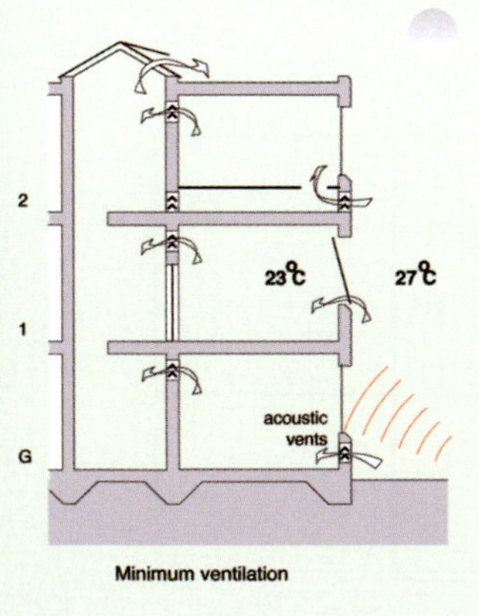

Night ventilation

The thermal mass of buildings can be used to soak up heat gains in the day, thus reducing peak temperatures, figure 10. But the heat has to be taken out of the building at some time. This is done most efficiently by maximising the ventilation rate at night. This technique can be so successful that the daytime temperatures in the building can be considerably lower than the peak outdoor temperature (up to 3 or 4°C). In this case it pays also to reduce the ventilation rate in the day when (and if) the outdoor temperature is above the indoor temperature.

In night ventilation strategies the following should be considered:

- Provide openings which can be left open at night, but maintain security.
- Consider how large volumes of air will flow through the building, from room to room and floor to floor.

- Thermal mass is only effective if it is coupled to the occupied space, and to the night ventilating air. Lightweight internal finishes to heavyweight buildings makes them behave as lightweight buildings – the benefit is lost. Consider exposing massive elements if covered – e.g. suspended ceilings.
- In lightweight buildings, consider adding thermal mass (e.g. new partitioning) or even phase change materials (see Tech Mon 1).

Note that refurbishment may involve the application of insulation internally to solid masonry walls. This will significantly reduce accessible thermal mass. If this has to take place, ensure that internal partitions, and if possible floor slabs, are heavyweight and not isolated.

Hybrid systems

Natural and mechanical ventilation need not be mutually exclusive. Obviously, certain spaces in a naturally ventilated building can be mechanically ventilated if they are internal or have high ventilation demands, such as toilets or kitchens.

Much fan power is wasted ventilating unoccupied or lightly occupied spaces. Demand control is where fans are only run when the air quality (as detected by CO_2 levels) is deemed unsatisfactory. CO_2 detecting controls are now cheap and reliable. Fans can be used to supplement natural air-flow in ducts and chimneys when the wind and buoyancy forces are too weak. This function also needs to be activated by a control system which detects a reduction of air-flow or a fall in air quality.

The arguments for and against hybrid systems are:

For – Passive systems would have to be oversized in order to cope with 'worst case scenarios – by accepting mechanical intervention, with appropriate controls, an optimum balance between energy efficiency and comfort can be struck.

Against – Necessitates the capital expense and maintenance of two systems

One technique where a small amount of mechanical power can be used with great effect is the use of ceiling or desk fans. These devices provide air movement but not fresh air. The air-movement can make a reduction in the effective temperature, i.e. as perceived by the occupants, of as much as 3°C. Although they improve thermal comfort, they do not improve air quality.

Summary conclusions

Natural ventilation can achieve high level of comfort, However, it requires considerable co-operation from the occupants and management, so it is important that both understand the principles and are aware of the problems.

Natural ventilation cannot provide such consistent and uniform conditions as mechanical systems. However, the positive side of this, is that there is growing evidence that variation in indoor conditions is tolerated and even enjoyed, provided the occupants are in control (see monograph "Adaptive thermal comfort standards and controls"). Other factors which have to be considered include the outdoor conditions, for example natural ventilation may be impractical in very noisy or polluted environments. However, it is observed that even in noisy urban environments, people will open windows, trading thermal comfort for traffic noise.

It is important to consider the effect of the whole package of refurbishment measures since they interact with on and other. For example, a building where air-conditioning had been identified as the only solution to comfort problems, might provide a satisfactory environment with natural ventilation, provided measures such as shading and fabric improvement were also carried out.

Hybrid systems should be considered where mechanical systems act as 'failsafe', activated only when air quality or thermal comfort falls below an acceptable level.

Prepared by Nick Baker the Martin Centre University of Cambridge

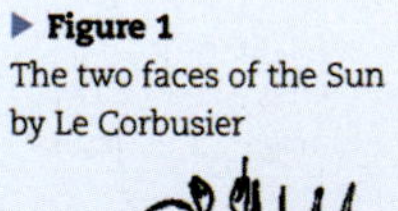

TECHNICAL**MONOGRAPH** 4

High performance daylighting - light and shade

This technical monograph is one of a set produced as part of the 'REVIVAL' project – an EU Energie Programme supported demonstration project of energy efficient and sustainable refurbishment of non-domestic buildings in Europe. The monographs explore some of the main energy and comfort issues which arose during the Design Forums held with each of the six sites. The four monographs are entitled:

- Thermal mass and phase change materials in buildings
- Adaptive thermal comfort standards and controls
- Natural ventilation strategies for refurbishment projects
- **High performance daylighting**

▶ **Figure 1**
The two faces of the Sun by Le Corbusier

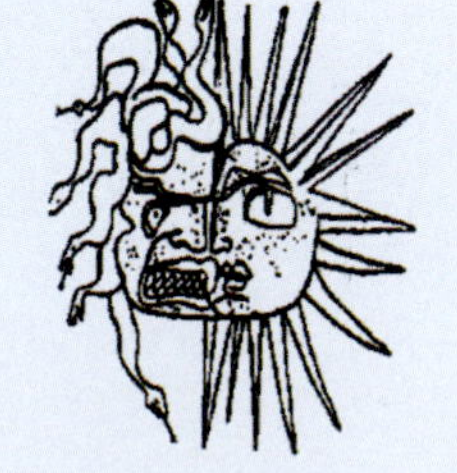

Refurbishment often involves the upgrading of glazing. This can range from straightforward replacement of the glazing material to the reduction or enlargement of the apertures. The installation of shading systems is also common, and again this can range from minor interventions such as roller-blinds, to large external structures. In the latter case, the design is often driven by architectural decisions rather than functional performance. However the daylighting will be affected, and in most cases, there will be an opportunity for improvement.

What is high performance daylighting?

We are all familiar with the problem facing the designer – large windows to let in lots of daylight and useful solar gains in winter, or small windows to conserve heat and prevent overheating from unwanted solar gains in summer. All too often, we observe in summer the 'blinds down - lights on' syndrome, and the presence of massive and costly shading structures which not only shade unwanted sun, but also cut out valuable daylight and obstruct views. It is also common practice to orientate the main glazed façade of buildings south (in the northern hemisphere) to maximise solar gain in winter.

Two recent technical developments have contributed to a solution but not fully solved it. Firstly, the U-value of modern glazing systems, using low-emissivity surfaces with inert gas filled cavities, can be as low as 1.0 $W/m^{2o}K$, 5 – 6 times better than single glazing. Secondly, 'high performance' glazing of this type may also have selective transmission to radiation that, by having lower transmittance in invisible part of the spectrum, reduces solar gain to a greater extent than the visible daylight.

However, this property is fixed, and clearly in the heating season, when thermal gains may be used to offset heating loads, this property is counter-productive. Given that most European buildings have both heating and [potentially] overheating seasons, how does the designer arrive at a good solution for year round performance?

Shading devices provide opportunities to both modulate and spatially re-direct incoming radiation. Furthermore, if movable, they can alter their function according to seasonal changes and even hourly sky conditions. This monograph sets out the technical principles for developing good solutions involving both high performance glazing materials and shading devices.

Glazing

The spectrum of daylight

The Sun is the original source of daylight; solar radiation falls on the outer atmosphere with an intensity of about 1.6 kW/m^2, where it is reflected, absorbed and scattered. This reduces the intensity of the direct beam to around 800W/m^2 by the time it reaches earth. About 50% of the energy lies in the visible region, the remainder being in the ultra-violet and infra-red spectrum, as shown in figure 2. In regions where clear skies do not predominate, useful daylight consists mainly of radiation scattered from clouds and the atmosphere, and will typically be about 50 – 150 W/m^2. Both absorption and scattering change the spectral content of the sunlight; particles of the atmosphere scatter shorter wavelengths (the blue end of the spectrum) preferentially. Thus diffuse daylight from the sky is slightly bluer than direct sunlight. The essential issue is however, that only the visible part of the spectrum provides

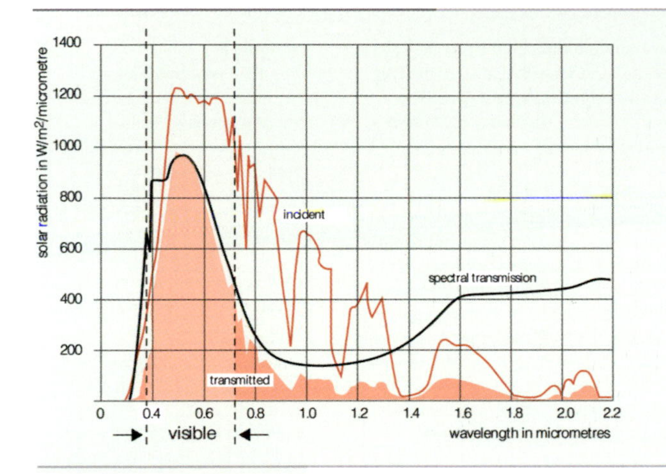

▲ **Figure 2**
The spectral distribution of sunlight. Only about half of the energy lies in the visible region. The spectral transmission of glass, and the resulting transmitted spectrum (shaded) is also shown.

daylight useful for visual tasks, whilst the whole spectrum, visible and invisible, contributes to heat gains when absorbed in the room.

The energy balance at glazing

Figure 3 shows the energy balance of the radiation incident on the glazing. The radiation falling on the glazing surface becomes three components – a reflected component I_r, a transmitted component I_t and an absorbed component I_a. For normal clear glazing, the reflected component is about 15% of the incident value but this value increases strongly when the angle of incidence increases beyond about 60°. The transmitted component is about 80% of the incident radiation, leaving 5% absorbed by the glazing.

The transmittance is the ratio of the transmitted component to the incidence component, often quoted as a percentage.

Two other types of glass are commonly encountered, tinted and reflective. Tinted or absorptive glass contains pigments to increase the absorption. This reduces the transmittance, typically from 40% to as low as 10%. The

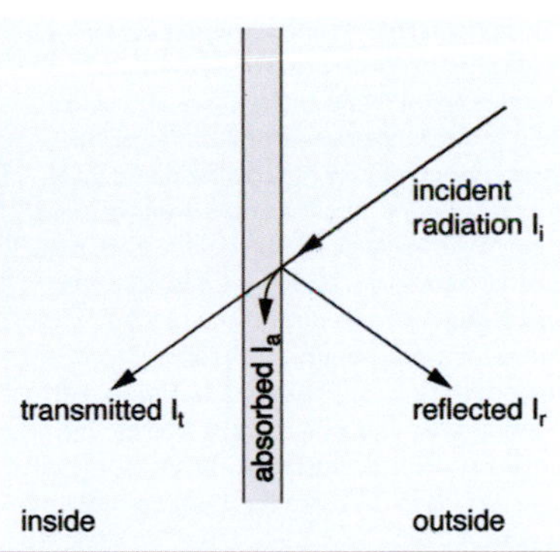

▲ **Figure 3**
Incident radiation is split into three components – transmitted, reflected and absorbed

absorbed energy heats up the glass and this heat is partly conveyed into the room and partly to the outside by radiation and convection.

Reflective glass has a thin metallic or semi-conducting coating which increases the reflected component, also reducing the transmission. However in this case, the energy is reflected away from the glass, not absorbed by it, and thus causes no heat gains to the room.

Glasses which have both reflective and absorptive properties have been available for about three decades. However, these materials reduced the visible part of the spectrum more than the non-visible. In spite of the glasses being marketed as having a beneficial environmental performance, there is no advantage in having a large area of low transmittance glass compared with a smaller area of normal glazing. Indeed, there is a disadvantage in that the glazing would have a much larger U-value than a well-insulated opaque envelope that could have replaced it.

High performance glazing products

Glasses which are referred to as "high performance" have the important property that the transmittance in the invisible spectrum (the infra-red and the ultra-violet) is significantly less than in the visible. Although the thermal gains due to the absorption of the visible light remains the same (for a given level of illuminance), gains from the invisible part of the spectrum are reduced. Thus the light can be regarded as "cooler", or put more scientifically, to be of *higher luminous efficacy*.

This is illustrated in the figure 4 which shows the heat gains to a 12m² room, for different glazing materials. In each case, the area of glazing is adjusted to give an average illuminance of 300 lux. The x-axis shows the ratio of total transmittance T_t to the visible transmittance T_v. For high performance glass, the ratio T_t/T_v is less than one. However, even if none of the invisible radiation were transmitted, this ratio could not be below 0.5, since about half of the thermal effect of solar radiation is due to visible radiation. This is shown as the theoretical limit on the graph. Materials with ratios greater than one actually worsen the situation, since, as the graph shows, more heat is generated in the room for the same amount of light.

This is also shown in figure 5 where the actual solar gains are calculated for a 16m² room. For the clear glazing the area has been set at 3.2 m² to provide a minimum daylight factor of 2%, a typical value for a well daylit room. The area has been adjusted to compensate for the reduced transmission in the other cases.

Energy re-distribution

It is the absorption of the solar radiation that causes the thermal gains, and this takes place on the surfaces of the room, as well as in the glazing itself. Figure 6 shows that the heat generated in the glazing is redistributed, partly inside, and partly outside, by convection and radiation. Within the room the process is complex and includes short-wave reflected radiation (visible, IR and UV) some of which 'escapes' back out through the glazing. However, most of the short wave IR and all of the long-wave IR is absorbed by normal glass. This, creates the diodic effect commonly known as the greenhouse effect.

Low-emissivity glasses (low-e)

Normal glass absorbs all long wave radiation (as a black surface does to visible radiation). It is also of high emissivity, that means that it emits almost the maximum amount of long-wave radiation possible for a given temperature. Low-e glazing has been coated with a very thin electrically conducting layer and this renders it reflective to long-wave radiation and by the same mechanism, reduces the emitted radiation, compared with normal glass.

The original use of low-e glass was to reduce the heat loss by radiation during the heating season, as shown in figure 7. It is usually used as part of double glazing, partly to protect the coated surface, and partly to ensure that there is still air adjacent to the low-e surface, to realise the maximum benefit

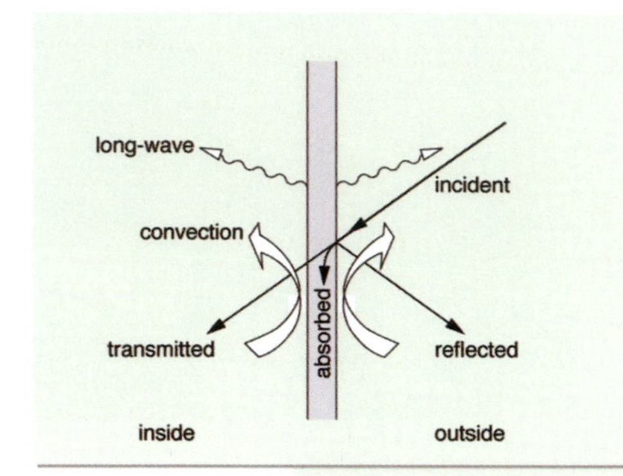

▲ **Figure 6**
Absorbed radiation hats up the glass. This heat is lost by longwave radiation and convection to the room and to outside.

▼ **Figure 4**
Solar heat gain plotted against the ratio of total transmittance to visible transmittance, for a given average illuminance of 300 lux

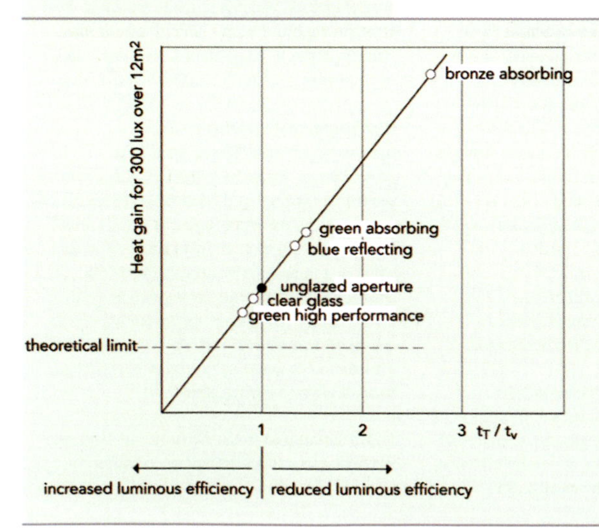

▼ **Figure 5**
The impact of glazing type on solar gain for a 4m x 4m sidelit room with a minimum Daylight Factor of 2%. The glazing area of 3.2 m² is adjusted to account for the reduced transmission of the tinted and high performance glass, to 7.1 and 4.9 m² respectively. The solar gain is calculated from: solar gain = 0 .75 x (area glaz) x $T_{t\,kW}$

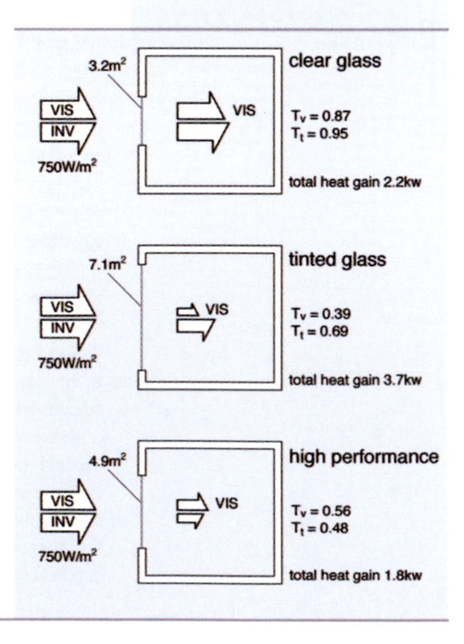

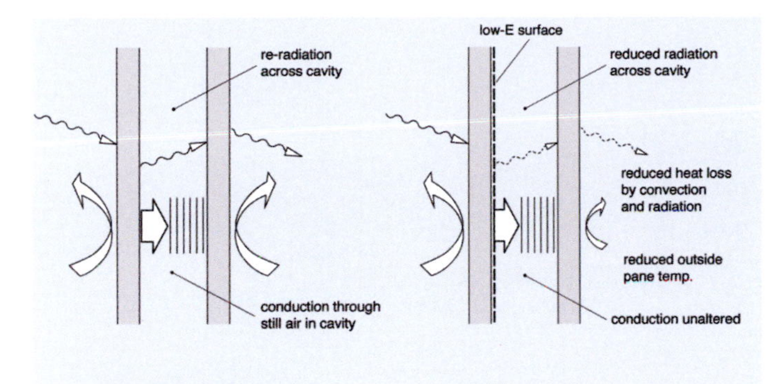

▲ **Figure 7**
The effect of a low emissivity coating on the inside of the cavity of double glazing. Heat transfer by long-wave radiation from the inner to outer pane is reduced

For example, the typical value of 2.8 for normal double glazing can be reduced to about 1.8 when a single internal surface is low-e. Convective heat transfer can be reduced by filling with heavy inert gas (argon) and this reduces the U-value further to about 1.0.

Recently, hard durable low-e coatings have been developed which can be applied to single glazing. This can reduce the U value from 6.0 to about 4.0 in still air.

Low-e surfaces can also contribute to reducing unwanted solar gains by reducing the heat transmitted from the outer pane to the inner pane by long wave radiation.

Dynamic transmittance, photo-, thermo, and electro-chromic glasses
The materials described previously have fixed properties. Thus they cannot respond to current conditions, and have to be specified as a compromise solution for long-term average conditions. One parameter which can be controlled dynamically is the transmittance. Photo-chromic materials reduce their transmittance as the light level increases. They are the most familiar, being in common use for spectacles. Costs still prevent their widespread use in buildings. However, their optical performance is not of the selective type, which means that glazing controlled in this way would suffer more solar gain than if controlled by conventional opaque shading devices.

Thermo-chromic materials, reducing their transmittance in response to an increase in temperature, are similar in that they are passive elements, responding to the agent (solar radiation), which needs to be controlled. However, in both cases, the materials are reactive, and cannot anticipate conditions, as well as having poor spectral transmittance characteristics.

Electro-chromatic materials are usually liquid crystal systems, similar in principle to those used for data display, responding to the application of an electric field. These offer the great advantage that they can be switched and modulated with control intelligence. However, they are not spectrally selective and so their thermal performance would not be particularly good.

Solar strategies
Materials with selective transmittance, as described above, are clearly aimed at reducing heat gains associated with useful daylight. In predominantly warm climates with mild winters, this is a priority. In most of these cases larger energy savings will be made by reducing cooling loads by shading, than reducing heating loads with solar gains. Many European climates however, also have a significant heating demand in winter. 'Solar strategies', where the building is designed to encourage winter solar gains, have long been seen as an important means of energy conservation.

The degree to which this is true is highly dependent on factors such as the detailed climate, the use pattern of the building and the presence of other sources of heat gain, the standard of insulation, and control strategy. It is probably fair to say that there are many cases where designers have tried too hard to maximise winter solar gain for only a modest reduction in heating load at the expense of other environmental problems; in particular overheating in summer.

Shading devices

Shading devices assist in the management of solar gains and useful daylight. Unlike glazing materials with fixed properties, shading devices – overhangs, fins, blinds, louvers etc, offer two ways in which they can respond to seasonal conditions.

Firstly, using simple geometry, fixed devices can intercept more solar radiation from higher angles than from lower angles, thereby approximately synchronising with the need for winter heating and the risk of summer overheating. Overhangs and fixed horizontal louvers, use this well known principle, which can be found in many examples of vernacular as well as contemporary architecture.

Secondly, moveable devices can be adjusted to modulate their transmittance – e.g. the opening and closing of louvers, or the raising and lowering of a low transmittance blind. The range of modulation can be large – in principle between 0 and 100%, as in the case of an opaque blind which can be fully retracted or fully deployed over a window. These categories and their significance to performance are described below.

Daylight redistribution

Shading devices have one more important property; due to the reflection of light from surfaces and/or the obscuration of light from specific directions, certain devices can improve the distribution of daylight in a room.
Sidelit rooms from plain apertures are very unevenly lit, as shown in figure 8. This means that in order that the minimum illuminance is sufficient, the part of the room near to the window has to be over-illuminated. At times of zero heating load this over-illumination serves no useful purpose, but will contribute to overheating and probably glare.

If the light distribution can be improved the total amount of light flux entering the room, (and the invisible radiation accompanying it), can be reduced. Thus the ideal shading device not only modulates the transmittance of the window, compensating for conditions of high radiation levels, but also improves the efficiency of distribution. Figure 8 shows how a light shelf reduces illuminance close to the window more than at the back of the room. Thus, the unwanted part of the incident radiation is reflected away reducing solar gain. Horizontal louvres can work in a similar way.

How can we assess performance?

If we consider the window, glazing and shading device as a *daylighting system*, then it is a useful concept to consider the *luminous efficacy* of this system – i.e.

Luminous efficacy =

$$\frac{\textbf{total useful light flux delivered to the work plane}}{\textbf{total radiation energy entering the room}}$$

Note that the definition of *useful* flux excludes illuminance above a datum value, i.e. a value that is considered to be sufficient for the function of the room.

The Luminous efficacy of daylight is about 110 lumens/Watt. This would be a theoretical maximum for a window system, i.e. providing an even illuminance at the datum value. In reality, a normal sidelit room, due to its non-uniformity could have a value as low as 30 lumens/Watt, placing it somewhere between tungsten and conventional florescent lighting. From an overheating point of view, we can see that there might be some justification for the "blinds down – lights on" mode of operation.

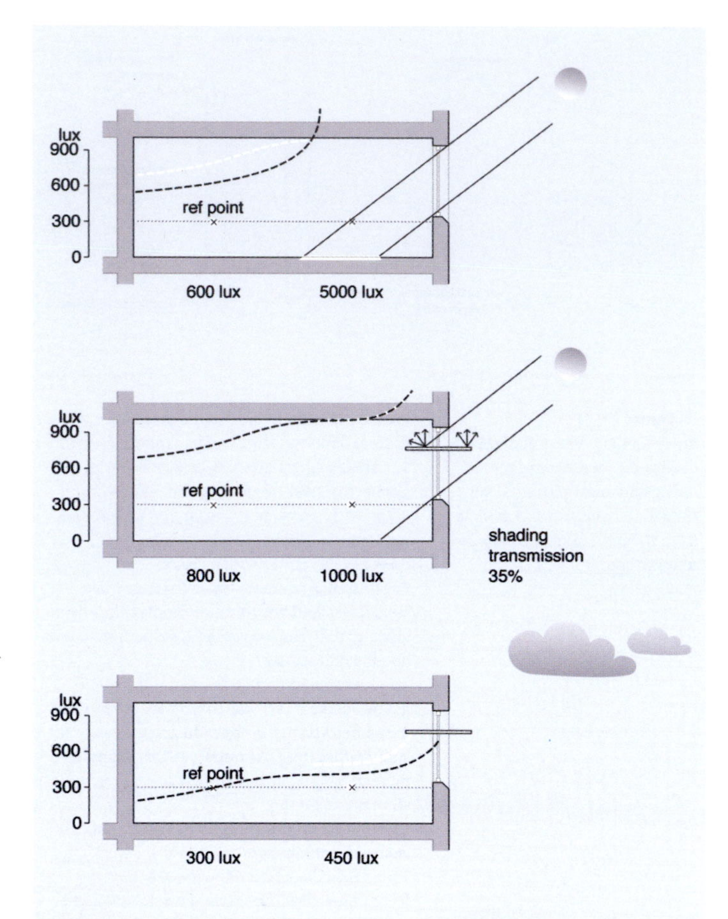

▲ **Figure 8**
Shows the non-uniformity of illuminance in a side-lit room (top) and the effect of a light-shelf in both sunny and cloudy conditions. The light-shelf reduces the over-illumination near the window.

Shading types – Dynamic performance

It is important to consider the performance of the system over different conditions – a solution that is optimum for a given combination of sky brightness and distribution may perform badly at other times. In particular, the ability to reduce the risk of overheating at times of plentiful solar radiation in summer, must be balanced against the ability to provide sufficient illuminance within the room, and possible useful solar gains, during the winter.

To assist in further understanding of this problem, it is useful to define four shading types –

Type 1 Retractable blinds, shutters and louvers.
This type does not affect the availability of useful daylight, provide they are under "sensible control" – i.e. they are deployed when there is a surplus of radiation, and retracted when there is minimal radiation. In this way they will not lead to an

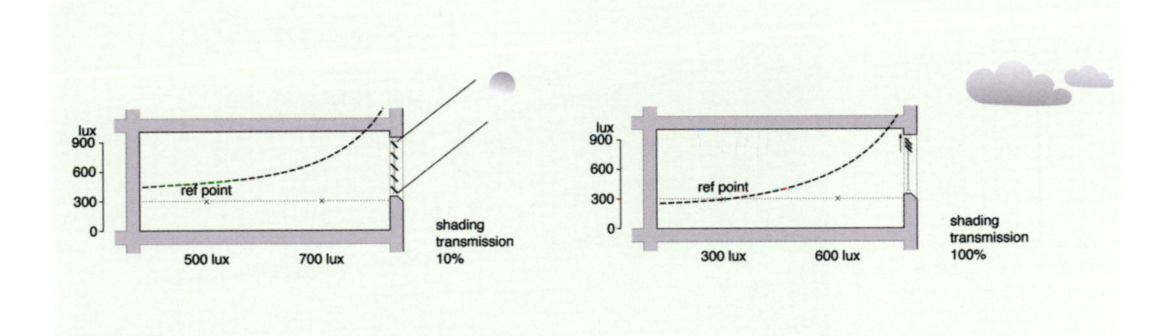

▲ **Figure 9**
Retractable shading should be deployed only when there is an excess of daylight illuminance, thus not affecting the limiting case when lights have to be switched on.

earlier switch on time for artificial lighting, and thus will not increase lighting energy.
On the other hand they will reduce overheating probability or cooling energy, as shown in figure 9.

Type 2 Fixed redistribution devices – overhangs, light shelves etc.
These devices reduce the total radiation passing through the window system, but owing to their being directionally selective, do not reduce the illuminance level at the switch-on reference point – i.e. they improve the luminous efficacy of the system, since they only reduce over illumination. This has already been shown in figure 8. Fixed devices of this type may reduce useful illumination compared with the unshaded aperture, but this can be compensated by increasing the aperture size.

Type 3 Reduced transmission selective (high performance) glazing.
This increases the luminous efficacy of the daylight (light to heat ratio). Slight reductions in illuminance at the reference point can be compensated by increased glazing areas, figure 10.

Type 4 Fixed obstructing screens and non-selective reduced transmission glazing (e.g. tinted, reflective, or fritted glass)
These devices reduce all radiation visible and non-visible, at all times, in the same proportion. Thus they will reduce the illumination at the reference point and lead to an earlier call for artificial light, as shown in figure 11. They offer no technical advantage over having a smaller unshaded aperture.

High performance daylighting in refurbishment projects

Glazing Replacement
Glazing replacement is commonly part of envelope refurbishment. This is most often to improve the thermal insulation performance, replacing single glazing with double or double low-e glazing. The main benefits of this will be in reduced heating load, but there should also be a significant improvement in comfort. This is due both to the reduced radiant losses to the glass surface, and the reduction of cold downdraughts, as illustrated in fig 13.

Low-e coatings are of two types (a) the hard ceramic coating and (b) the soft metallic coating. The hard coating has a slightly poorer performance than the soft coating, but it has

▼ **Figure 10**
Low-e glazing in the refurbished classrooms at Chevrollier

▼ **Figure 11**
Fixed screens and reduced transmission glazing (tinted, reflective or fritted) reduce the illuminance throughout the room and therefore increase the need for artificial lighting

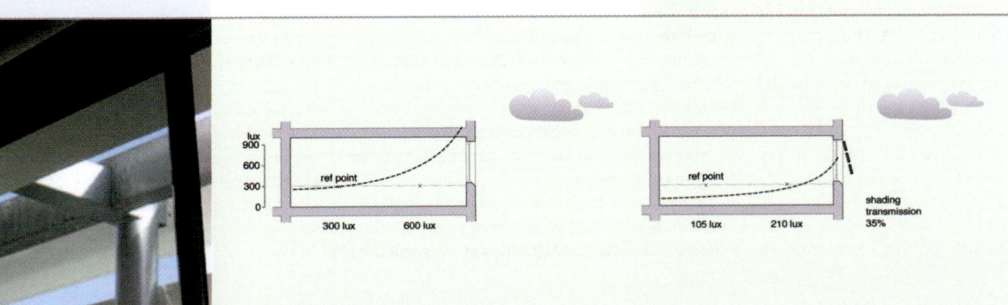

▶ **Figure 12**
Fixed external louvres may improve light uniformity but interfere with view out

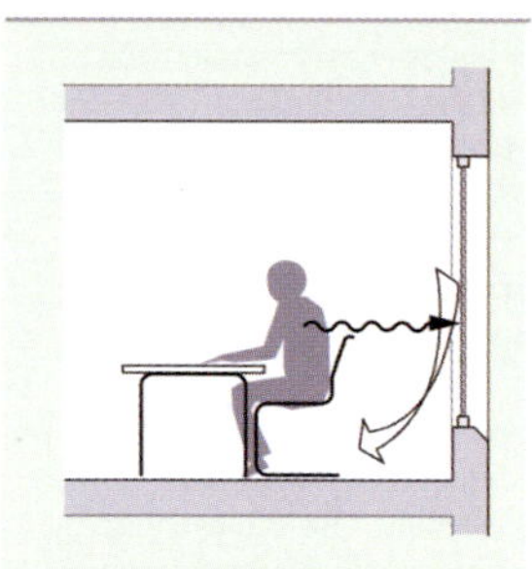

▲ **Figure 13**
Cold radiation and downdraughts are reduced by the use of low-e coatings thereby increasing comfort as well as reducing heat loss

the advantage that it is durable enough to be unprotected, i.e. it can be used in single glazing. This may have real advantages in the refurbishment of historic buildings where the original framing (glazing bars) is to be retained. Although, the reduction of U-value from 6 W/m^{2o}K to 4 W/m^{2o}K is much less than the 2 W/m^{2o}K achievable with low-e double glazing, this 33% reduction could be significant in a building with large areas of glazing.

Where the retrofitting of double glazing is possible, then the soft coating can be used. Units with argon filled cavities have U-values, as low as 1.2 W/m^{2o}K.

When assessing the thermal benefit of replacement double glazing, the thermal performance of the whole glazed envelope, i.e. including the glazing bars, must be considered. Framing systems of steel or aluminium without thermal breaks, and with small panes and therefore a larger area of framing, can have much reduced performance due to conduction through the framing itself. For example an aluminium frame with no thermal break and occupying only 10% of the aperture will increase the U-value of the glazing from 1.5 to 2.2 W/m^{2o}K.

Modifying apertures
Many buildings from the 50s onwards, that are now being considered for refurbishment, have large areas of glazing. In many cases this is single glazing, and would thus be possible candidates for glazing replacement. If aesthetic considerations allow it, a reduction of the aperture area could be considered. This will have several advantages – reducing heat loss and unwanted solar gain, reducing glazing costs, reducing shading costs, and in some cases supporting interior remodelling.
Since the new opaque envelope can have U-values as low as 0.2 W/m^{2o}K, the impact on the average U-value can be significant. For example, if the 70% double glazing of a façade is reduced to 35%, with an opaque material with U-value of 0.2 W/m^{2o}K, the average U-value of the original aperture would be reduced from 2.8 W/m^{2o}K to 1.5 W/m^{2o}K.

Although 35% glazing can be shown to provide adequate daylighting for rooms of up to 6m deep, care has to be taken with how the distribution of glazing is altered. General rules of thumb are given below for spaces with a 3m floor to ceiling height–

- Sill heights should not be raised above 1m from floor level.
- Glazing in the upper part of the wall is more effective than the lower (except when there are deep overhangs)
- Horizontal distance between glazed areas should not exceed 3m (or 2m from crosswalls in cellular offices
- There may be a case for splitting the viewing and daylighting function

There may be cases for increasing the glazed area, where daylighting is demonstrably inadequate. However, other causes for poor daylight performance should be eliminated first – e.g. low transmission of glass, obstruction due to framing or poorly designed shading devices, low reflectance of interior surfaces, or internal obstructions.

There may be cases for changing the distribution of glazing by making new apertures. Single-storey buildings and top floors offer a good opportunity for this. Small areas of rooflighting over a deep plan can make a dramatic improvement in daylight distribution, and therefore, on the luminous efficacy of the system. However, in warm climates, rooflights should never be un-shaded horizontal glazed apertures, but always structures with apertures sloping away from the equator.

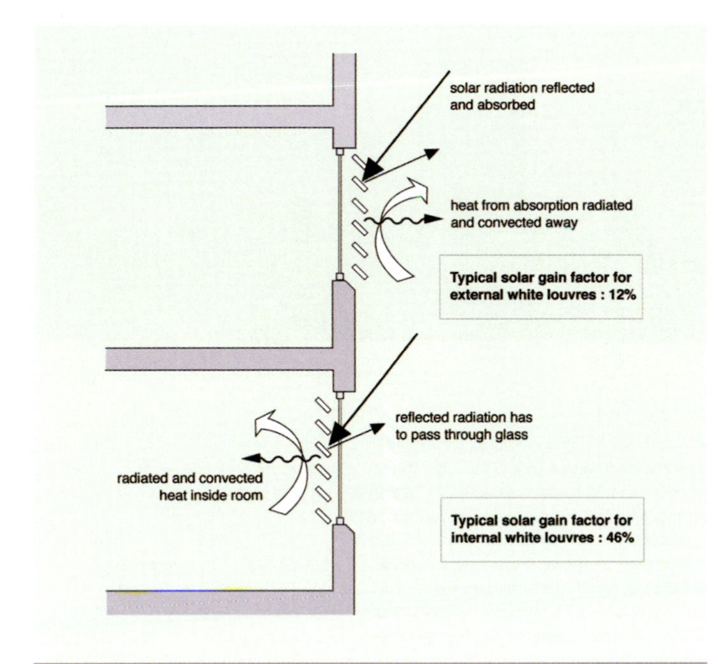

▲ **Figure 14**
External versus internal shading

Shading options
Many buildings lack shading. This may either be due to deliberate design intention, i.e. that the shading devices have been seen by the architect as an intrusion on the design concept, or due to economies. In some cases they simply weren't considered due to poor understanding of the problem.

Many of the buildings where the architectural specification has not included shading provision, will have had shading retro-fitted at some stage. Most commonly this will be internal curtains, blinds or louvres, but at worst, pieces of paper stuck onto the glass by desperate occupants. The installations are often poorly integrated with the window design, obstructing controls and interfering with ventilation, and are usually poorly maintained. These experiences give shading a poor reputation.

These notes are intended to assist in strategic decisions about shading. Detailed design and specification should use analytical design tools to assess the performance quantitatively.

Options for shading fall into four categories – **External, Internal, Inter-pane and Integrated.**

It is well known that the thermal performance of external shading is superior to internal shading, since re-emitted heat is lost to the outside, rather than into the room, as shown in figure 14. However, internal shading is usually cheaper and easier to control.

Interpane shading refers to devices held between panes, or in the cavity of a 'double skin' envelope. These systems have the advantage that the shading device is protected from weather, dust, and mechanical damage. It is important that the design of the cavity permits adequate ventilation to the shading elements.

Integrated refers to devices such as lightshelves and prismatic systems which explicitly address the daylight distribution function as well as selective shading.

External Shading
External shading broadly falls into the following types –

- Overhangs (fixed or retractable)
- Louvres (fixed, adjustable, retractable)
- Fins (fixed, adjustable)
- Blinds (retractable)
- Perforated screens (fixed)

fixed – fixed

adjustable – remains in position but radiation transmission can be modulated e.g. altering angle of louvre.

retractable – can be removed completely from the aperture

A summary of their properties is given in the table below. When using the table please note the following.

Nat vent limiting conditions refers to situations requiring the shading to be deployed, i.e. when there is an overheating risk.

Daylight limiting conditions refers to minimal daylight availability when there would be no requirement for shading.

Adjustable louvres and fins will have poor view performance and poor natural ventilation performance, if completely closed.

Orientation ***all*** implies that performance is not orientation-sensitive, although in general, there would be no demand for shading on facades orientated 360 +/- 45.

Figure 15 Office in Athens where internal louvre interfere with the opening windows and ventilation.

Internal shading. This is often the choice for retrofit. Options are limited to louvres (Venetian blinds) and roller blinds of translucent material.

Horizontal louvres offer some possibility for improved light distribution by using a combination of high and low reflectance finishes. If the upper surface is light, and the lower surface dark, light will tend to be directed upwards towards the ceiling, thereby improving the illumination at the back of the room. Some, proprietary systems use specially shaped louvres with specular (mirror) reflecting surfaces, in order to further improve the performance.

Most internal louvre systems are retractable, allowing maximum daylight transmission at times of limiting daylight availability. This also permits cleaning. Whilst open louvres can allow a good flow of air, they should be anchored top and bottom to prevent movement and noise. However, access to window opening controls must be provided. These conflicts often result in low cost simple installations being of poor functional quality, as shown in figure 15.

Shading type	Orientation degrees from North	View	Nat vent (in limiting conditions)	Daylight (in limiting conditions)	Seasonal response	Modulation	Notes
Overhangs							
fixed	180 +/- 30	good	good	medium	medium	none	e.g. canvas awnings +
retract	180 +/- 30	good	good	good	good	good	adjustable geometry
Louvres							
fixed	180 +/- 30	med – poor	good	medium	medium	none	View influenced by blade
adjust	all	med – poor	good	good	good	good	module size & geom.
retract.	180 +/- 30	med/good	good	good	good	medium	'good' applies to when
retract.+ adjust	all	med/good	good	good	good	good	retracted
Fins (vertical)							
fixed	90, 270 +/- 20	med – poor	good	medium	medium	none	View influenced by blade
adjust	90, 270 +/- 45	med	good	good	good	good	module size & geom.
Blinds							
retract	all	poor/good	poor	good	good	medium	'good' applies to when retracted
Screens fixed	all	poor	med - poor	poor	poor	none	not recommended

Roller blinds are another common retrofit solution due to their low cost and ease of installation. Consideration must be given to the optical properties of the fabric as indicated in figure 16.

Roller blind fabrics are optically diffusing, which means that light is re-emitted in all directions. This means that blinds of moderately high light transmission could become glare sources themselves.

When roller blinds are used in conjunction with overhangs, consideration should be given to positioning them so that they deploy upwards – i.e. with the roller at cill level. This means that sunlight striking the lower part of the window beneath the overhang will be intercepted, whilst allowing light and view through the upper part. Alternatively, the lower part of the window can be shaded separately.

With both louvre and roller blinds used internally, it must be noted that any absorbed radiation will result in heat released within the room. This means that generally blinds of high reflectance (lighter colour) will reflect a greater proportion of the unwanted visible radiation out through the glazing, thereby increasing the luminous efficacy of the system.

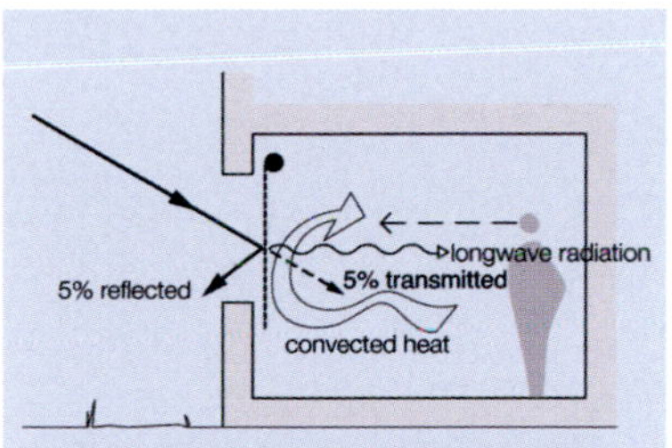

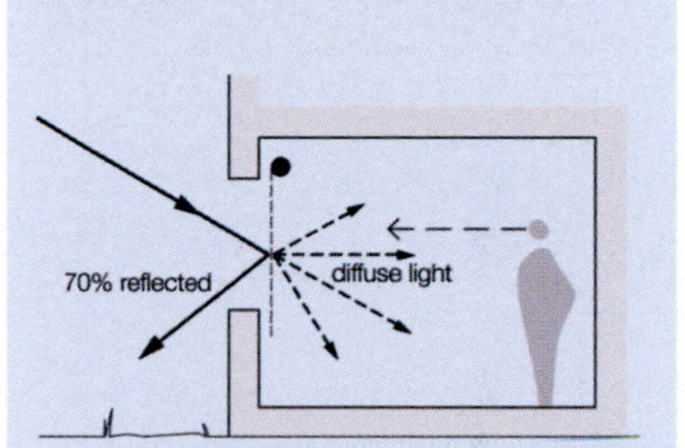

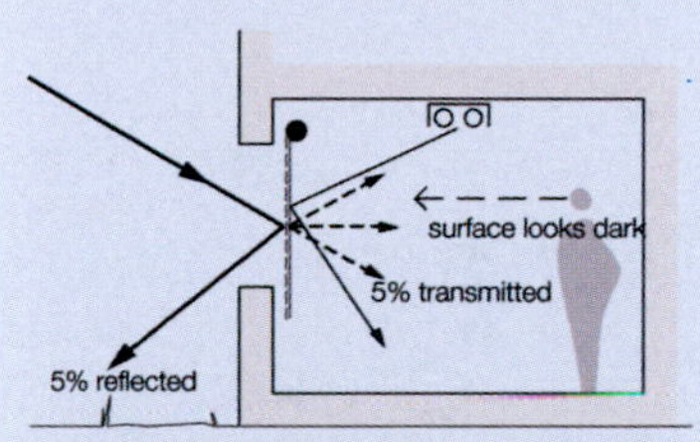

▲ **Figure 16**
The performance of various types of blind materials

Summary Conclusions

High Performance Daylighting is an integrated approach that takes all the functions of the window system into account. By adopting selective devices (spectral and spatial) and dynamic devices (e.g. retractable and adjustable louvers), the integrated year round energy performance can be optimised.

Recently developed high performance glazing products increase the luminous efficacy of the daylight, by transmitting less of the invisible radiation. Thus cooling loads can be reduced without increased artificial lighting energy. However, many conventional glazing products such as tinted, reflective and fritted glazing materials do not have this property.

The effective luminous efficacy the window system can be increased by improving the illuminance distribution, i.e. reducing unwanted over-illumination near the window without reducing it at the back of the room.

The time-integrated luminous efficacy can be improved by adjustable shading which responds to the prevailing sky conditions.

In most cases, large improvements can be made with glazing and shading products without necessitating major structural alterations. This makes high performance daylighting an appropriate objective for refurbishment projects.

Prepared by Nick Baker The Martin Centre University of Cambridge

Index

A

absorption chillers 77
absorptive glass 49, 50–1
acoustics 68–9, 100
actuators 81
adaptive controls 84–6, 114, 116–17
adaptive opportunities 7–8
adjustable shading systems 58–9
advanced natural ventilation 9
air-conditioning 7–9, 72–3
  case studies 111, 114–16, 120–1
  deep-plan buildings 5
  energy performance 15–17
  hybrid/mixed mode systems 86–7
air heating systems 72–3
air quality 66, 97–8, 100, 105, 125–6
air requirements 116
air temperatures 74–5, 125–6
The Albatros, Den Helder 6, 54, 93–100
aluminium coatings 60
aluminium frames 53, 94–5
angles of incidence 52
aperture modifications, windows 55–6
architectural integration 18–19
artificial lighting 11, 77–80
  The Albatros 98
  Athens Ministry of Finance 124, 127
  daylight detection 83
  Lycée Chevrollier 103–5, 108–9
  occupancy detection 81–2
  see also lighting
Athens Ministry of Finance 123–9
atria 63–70, 102, 105–6
  see also double skins
attic spaces 39–40, 44

B

BEMS *see* building energy management systems
biomass heating 89–90
blinds 58, 59–60, 86
boilers 71, 76
break-even point, rebuild 4
BREEAM (Building Research Establishment Environmental Assessment Method) 22
bridging, cold bridges 28, 32–4, 42, 45, 102
budget constraints 13
building energy management systems (BEMS) 83–4, 95–6, 99, 104–6, 124–5, 132–3
Building Regulations, UK 21–2
Building Research Establishment Environmental Assessment Method (BREEAM) 22
buildings as sub-systems 4–5
bulk materials 3
  see also concrete

C

carbon dioxide *see* CO_2 emissions
Carbon Reduction Commitment (CRC) 22
caretaker controls 86
cavities (roofs) 39–40, 44
cavity walls 35–7
ceiling fans 124
cellular offices 83

central control systems 80–1
change of use 5–7, 93
chillers 77, 87, 113–14
CHP *see* combined heat and power systems
climatic factors
  atria 65
  natural ventilation 18
CO_2 emissions 3–23
  The Albatros 94–5, 97–8
  Athens Ministry of Finance 125–6, 128
  Daneshill House 114, 116–17, 119–21
  fans/pumps 76
  heat distribution systems 72
  Lycée Chevrollier 104–5, 107–10
  Meyer Hospital 134
  office buildings 5
  refrigeration 77
coated glass 49–50
Coefficients of Performance (COPs) 77
cold bridges 28, 32–4, 42, 45, 102
combined heat and power (CHP) systems 71, 77
comfort
  The Albatros 100
  Athens Ministry of Finance 125–6, 128–9
  heat emitters 74–6
  hybrid/mixed mode systems 87
  Lycée Chevrollier 102, 108–9
  Meyer Hospital 134
  standards 7–9
commissioned buildings 110
  see also newbuild
composite walls 35, 36–7
concrete 3, 9, 111–13
condensation 34–5, 36–7, 40–1, 43, 76
condensing boilers 71, 76
conduction energy transfer 48, 50
consumption *see* energy performance; gas consumption
control systems 71–87
  The Albatros 98–9
  Athens Ministry of Finance 123–9
  Daneshill House 114–17
  Lycée Chevrollier 104, 107
  Meyer Hospital 134
convection energy transfer 48, 50
convector heaters 73–4
CoolDeck system 112–14, 120
  see also cold bridges
cooling systems 76–7
  Athens Ministry of Finance 124, 127–8
  Daneshill House 112–14, 116, 120
  evaporative 45–6
  hybrid/mixed mode systems 86–7
  underfloor 27–8
  see also air-conditioning; natural ventilation
coolth emitters 76
COPs (Coefficients of Performance) 77
cost effectiveness
  mechanical services 71
  renewable energy 89
  see also economic factors
County Hall, London 11
CRC (Carbon Reduction Commitment) 22
curtain walling 51

D

dampers, cooling systems 116
Daneshill House, Stevenage 87, 111–21
daylight detection systems 82–3
daylight factors (DFs) 82–4, 98–9, 103
daylight redistribution 56–7, 59
daylighting 10–13, 56–7, 59, 77, 82–4
  The Albatros 98–9, 100
  Athens Ministry of Finance 124, 127
  atria 67–8
  Daneshill House 111
  glazing materials 47, 49–50, 55–6
  high performance daylighting 60–1
  Lycée Chevrollier 101–4, 108–9
  Meyer Hospital 133–4
daytime temperatures, atria 65–6
DECs (Display Energy Certificates) 22

deep-plan buildings 5, 11–12
demand controls 79–81, 124–5
dewpoint profiles 34, 36–7, 41
DFs *see* daylight factors
digital controls 81
dimming controls, lighting 82, 85, 98–9
direct expansion (DX) units 113–14
Display Energy Certificates (DECs) 22
double glazing 48–9, 55
double skins 58, 63–70, 93–8
ductwork 72
DX (direct expansion units) 113–14

E

earth constructions 31
eco-communities 19–20
economic factors 3–4
  see also cost effectiveness
EEAS (Energy Efficiency Accreditation Scheme) 22–3
electric heaters 75
electrical energy
  The Albatros 99–100
  CO_2 emissions 76, 77
  Daneshill House 117, 120–1
  Lycée Chevrollier 102, 107
embodied CO_2 3, 94–5, 109
emissions *see* CO_2 emissions
energy conservation 20, 63–7, 68, 83
Energy Efficiency Accreditation Scheme (EEAS) 22–3
energy performance
  The Albatros 98–9
  Athens Ministry of Finance 124–8
  atria 63–7
  building/services/occupants system 4–5
  change of use 6–7
  Daneshill House 112–19, 121
  daylight detection 83
  double skins 68
  eco-interaction 20
  low emission refurbishment 3
  LT Europe software 15–17
  Lycée Chevrollier 107–8
  Meyer Hospital 134
  quantifying 14–18
Energy Performance Certificates (EPCs) 21
Energy Performance of Buildings Directive (EPBD) 20–2
energy supply, eco-interaction 20
energy transfer, glazing 47–50
engineering integration 19
environmental benefits 3–4, 13–14
environmental comfort standards 7–9
  see also comfort
environmental conditioning 79–80
environmental regulations 20–3
environmental strategies *see* strategies
EPBD (Energy Performance of Buildings Directive) 20–2
EPCs (Energy Performance Certificates) 21
European Union (EU) directives 20–2
evaporative cooling 45–6
external insulation, walls 31–3
external shading systems 57, 58–9

F

fan coils 74, 76
fans 72–4, 76–7, 99, 112, 124
feedback controls 85–6, 87
filters, air-conditioning 73
'fin effect' 53
fin shading systems 58
fixed shading systems 58–9
flat roofs 40–1
floors 27–9, 102–4
flow resistance, heat distribution 72–3
framing systems
  double skins 94–5
  wall insulation 31, 33
  windows 51–5
free cooling systems 116

G

gas consumption 107, 118–21
glazing 47–51, 53–6, 60
  The Albatros 93–5
  Athens Ministry of Finance 123–4
  atria 67–8
  double skins 63, 69
  lighting systems 11–12, 16–17
  over-glazed buildings 7–8
  see also windows
global lighting efficacy 78
green roofs 45–6
greenhouse, Meyer Hospital 131, 132–4
ground floors 27–9
ground sourced refrigeration 77

H

heat distribution 72–3
heat emitters 73–6
heat exchangers 71
heat gains
  glazing 50–1
  natural ventilation 9–10
heat loads
  atria 65–6
  lighting 16
heat transfer/transmission 36, 47–50
heating systems 71–6
  The Albatros 94, 96–100
  Athens Ministry of Finance 124, 127–8
  Daneshill House 111, 118–21
  local control 80
  Lycée Chevrollier 105, 107
  renewable energy 89–90
  underfloor 27–8, 74, 111
high performance daylighting 60–1
high performance glazing 50–1, 53–4
historic buildings 54–5, 131–2
horizontal louvres 59
hospital buildings 131–4
hybrid systems 8, 9, 86–7

I

illuminance levels/distribution 79, 82–3, 85
indoor air quality *see* air quality
infiltration rates 9
infrared (IR) light 44, 50
inner leaf, cavity walls 37
insulation
  cavity walls 35–7
  floors 27, 28–9
  Lycée Chevrollier 101–3, 107
  Meyer Hospital 131–2
  roofs 39–46
  solid floors 27
  solid walls 31–4
  suspended floors 28–9
  thermal response 29
  walls 31–4, 35–7
integrated newbuild/refurbishment 18–19
integrated shading systems 57–8
intelligent skins 63, 68
intermediate floors 29
internal insulation, walls 33
internal shading systems 57, 59–60
internally reflected components (IRCs) 103
interpane shading systems 57–8
interstitial condensation 34–5, 36–7, 40–1, 43
IR (infrared) light 44, 50
IRCs (internally reflected components) 103

K

Kalamazoo building, Birmingham 12

L

lamps 78–9
LEDs (light emitting diodes) 116–17
legislation 20–3
life-cycle performance 94
light emitting diodes (LEDs) 116–17
light obstructions 52–3
lighting

The Albatros 93–4, 98–9, 100
Athens Ministry of Finance 124, 127–8
controls 81–5, 104, 114–17
Daneshill House 111, 114–17
energy performance 16–17
installations 77–80
Lycée Chevrollier 101–5, 107–9
Meyer Hospital 134
see also daylight...
lightshelves 56–8
lightweight building comfort standards 7
lightweight walls 35, 36–7
load-bearing insulation 27
local control systems 79–80, 81
louvred shading systems 56–8, 59, 103, 132
low emission refurbishment 3–23, 44, 49–50
low-emissivity coatings 49–50
low-emissivity membranes 44
LT Europe software 14–17
luminaires 78–9, 81–2, 84, 103–4, 114
see also lighting
luminous efficacy 50–1, 60, 77–8
Lycée Chevrollier, Angers 9, 18, 101–10

M

masonry walls 35, 36–7
mechanical services/controls 71–87, 114, 123–9
Meyer Hospital, Florence 5, 56, 131–4
Ministry of Finance Offices, Athens 123–9
mixed mode systems 8, 9, 86–7

N

natural ventilation 8, 9–11, 15, 17, 18, 59
newbuild 3–4, 18–19
see also commissioned buildings
night ventilation 9–10, 17, 104, 112–13, 124

O

occupancy demand matching 79–81, 124–5
occupancy density
The Albatros 97–8
Daneshill House 119–20
occupancy detection lighting controls 81–2, 104, 114
occupant controls 84–6
occupants as sub-systems 4–5
office buildings
case studies 93–100, 111–21, 123–9
change of use 6, 93
CO_2 emissions 5, 72
comfort standards 7
lighting 79, 83
opaque surface acoustics 69
open plan buildings 83, 86
opportunities, adaptive 7–8
optical transmittance 47
outer leaf, cavity walls 37
over-glazed buildings 7–8
overhang shading systems 58, 60
overheating performance
Daneshill House 111, 121
LT Europe software 15–17
Lycée Chevrollier 101–2
Meyer Hospital 133
walls 31
windows 50–1

P

PAC (partially air-conditioned) buildings 9
paints, roofs 44
partially air-conditioned (PAC) buildings 9
passive systems 8–12, 86–7
PCM (phase change materials) 112–14
perforated screens 58
performance *see* energy performance; overheating performance; thermal performance
permanent supplementary artificial lighting

(PSALi) 11
personal benefits 13–14
phase change materials (PCM) 112–14
photo-sensitive detectors 82, 86, 104, 114, 116
photovoltaic (PV) applications 89–90, 102, 106, 127, 133
pipework, heat distribution 72
pitched roofs 39, 40–1, 43
planning policies, UK 22
plants
  atria 68, 69
  green roofs 45–6
  shading systems 132
polyurethane insulation 43
positioning heat emitters 75
pre-cast concrete 9
principles 1–23
prioritizing options 13–18
protectivity, atria 64–5, 70
PSALi (permanent supplementary artificial lighting) 11
pumps 76–7
PV *see* photovoltaic applications

Q

quantification of energy benefits 14–18

R

radiation transmission 49–50
radiators 73–5
raised floors 27
rammed earth constructions 31
RE (renewable energy) 89–90, 93
rebuild vs refurbishment 3–4
recycling provisions 106
reflectance, roofs 43–4
reflective glass 49, 50
refrigeration 77
regulations 20–3
renewable energy (RE) 89–90, 93
retractable shading systems 58–9
REVIVAL buildings 5–6, 9, 56, 59, 86
rigid insulation 31, 33
roller blinds 59–60
roof ponds 45, 46
rooflighting 10, 56
roofs 39–46, 131

S

SBS (sick building syndrome) 11–12
school buildings 63, 101–10
screed finish, floors 27
second skins *see* double skins
secondary glazing 53–4
separating walls, atria 65–6, 68
service CO_2 94–5
service systems 4–5, 71–87, 114, 123–9
shading factors 51
shading systems
  adaptive control 86
  The Albatros 98
  Athens Ministry of Finance 123–4, 127
  atria 67–8
  Lycée Chevrollier 102, 103
  Meyer Hospital 132–3
  windows 51, 56–60
SHF (Solar Heat Fraction) 43
sick building syndrome (SBS) 11–12
side-lit rooms
  daylight factors 83–4
  shading systems 56–7
sizing heat emitters 75–6
socio-economic factors 3–4
solar gains
  roofs 39, 43–4
  solid walls 32
  windows 50–1, 56–7
Solar Heat Gain Factors 43
solar panels 89
  see also photovoltaic applications
solar water heating 118–20

solarity, atria 65, 69–70
solid ground floors 27–8
solid roofs 39, 41–3
solid walls 31–5, 131–2
sound insulation 69
space use efficiency 119
spatial mix systems 87
standards, comfort 7–9
steel 3
Stevenage 'new town' 111
strategies
  The Albatros 93–4
  Athens Ministry of Finance 123
  Daneshill House 111
  low emission refurbishment 3–23
  Lycée Chevrollier 101–2
  Meyer Hospital 131–4
stratification of temperature 66–7
structural deck, solid roofs 42–3
summer performance
  The Albatros 96, 100
  Athens Ministry of Finance 125–6
  atria 67
  Lycée Chevrollier 105–6, 108–9
support systems, windows 51–5
surface reflectance 43–4
surface resistance/conductance 48
suspended ground floors 28–9
systems concept 4–5

T

task lighting 79
temperature profiles
  Athens Ministry of Finance 125–6
  atria 64, 65–7
  cavity walls 36–7
  Daneshill House 113–14
  double skins 96–7
  heat emitters 74–6
  Lycée Chevrollier 104–5
thermal mass
  cooling systems 112
  roofs 44–6
  solid walls 32–3
thermal performance
  The Albatros 95–6, 100
  Athens Ministry of Finance 125–6
  atria 64–7
  double skins 68
  floor insulation 29
  glazing materials 47–50
  Lycée Chevrollier 101–2, 104–6, 108–9
  roofs 43–5
  wall insulation 32–3, 37
  window frames 53–5
thermal transmittance 47–50
  see also heat transfer/transmission
thermostatic controls 86, 107
timber frames
  walls 31
  windows 54–5
time-programming controls 80
tinted glass 49, 50–1
toilet use 99–100
transparent buildings 12

U

U-values 47–50
UF (utilization factor) 78
UK legislation 21–2
ultraviolet (UV) light 50
underfloor cooling 27–8
underfloor heating 27–8, 74–6, 111
'upside-down' roofs 41–2
UPVC-framed windows 54
urban reflectance 44
urban renewal 19–20
utilization factor (UF) 78
UV (ultraviolet) light 50

V

vapour checks
  roofs 41
  walls 34–5, 36–7
vapour permeability, roofs 40, 41, 43
vegetation
  atria 68, 69
  green roofs 45–6
  shading systems 132
ventilation 8–11, 15, 17–18, 59
  The Albatros 93–100
  Athens Ministry of Finance 124–6
  atria 65–7, 69
  cavity walls 35–6
  Daneshill House 112–13
  fans 76
  local control 80
  Lycée Chevrollier 102, 104–6, 108–9
  Meyer Hospital 132–3
  roofs 40, 41
  see also air-conditioning
visible spectrum 49–50
voids, roofs with 39, 40–1
voluntary schemes 22–3

W

Wallesey School 63
walls 31–7, 131–2
waste management 106
water heating systems 72, 118–21
water pumps 76–7
water surface, roof ponds 46
water temperatures 75–6
water vapour *see* vapour...
watering systems, plants 68
waterproof membranes 41–2
Wauquez Department Store 10
weathering function, roofs 39
wet cooling towers 77
wind direction, ventilation 98
wind turbines 89
windows 10–12, 47–61, 104–5, 131–2
winter performance
  The Albatros 95–6, 100
  Athens Ministry of Finance 126
  atria 65–7
  Lycée Chevrollier105 108–9
wireless digital controls 81
wood *see* timber frames

Z

zoning systems 72, 80–1, 83, 85